Climate Change Science

Written by an established climate change scientist, this book introduces readers to cutting-edge climate change science. Unlike many books on the topic that devote themselves to recent events, this volume provides a historical context and describes early research results as well as key modern scientific findings. It explains how the climate change issue has developed over many decades, how the science has progressed, how diplomacy has (so far) proven unable to find a means of limiting global emissions of heat-trapping substances, and how the forecast for future climate change has become more worrisome. A scientific or mathematical background is not necessary to read this book, which includes no equations, jargon, complex charts or graphs, or quantitative science at all. Anyone who can read a newspaper will understand this book. It is ideal for introductory courses on climate change, especially for nonscience major students.

RICHARD C. J. SOMERVILLE is Distinguished Professor Emeritus at Scripps Institution of Oceanography, University of California, San Diego. He is a fellow of three scientific societies: the American Association for the Advancement of Science (AAAS), the American Geophysical Union (AGU), and the American Meteorological Society (AMS). He has received awards from the AMS for both his research and his popular book *The Forgiving Air: Understanding Environmental Change* (1996, University of California Press). From the AGU, he has received two major honors: the Climate Communication Prize (2015) and the Ambassador Award (2017). Richard is a coauthor (with Sam Shen) of *Climate Mathematics: Theory and Applications* (2019, Cambridge University Press) and also a coeditor of *The Development of Atmospheric General Circulation Models: Complexity, Synthesis and Computation* (2018, Cambridge University Press).

This succinct collection distils the insights of one of the world's foremost climate scientists. Clear, engaging, and deeply informative, *Climate Change Science* is an illuminating guide for readers seeking clarity – and a much-needed response to the ongoing assault on climate science.

Professor Kerry A. Emanuel, Massachusetts Institute of Technology

At a time when misinformation and disinformation are polluting the discourse related to climate change, Somerville's book is a beacon of sound science written in an accessible manner. The "So-What's?" are clear.

Professor Marshall Shepherd, University of Georgia

Richard Somerville's *Climate Change Science: An Essential Reader* provides a masterclass in climate change communication, delivered by someone who has spent decades working in the field. The book is excellently written, thought-provoking, and informative. A highly enjoyable read that does not require you to have a background in mathematics or science before you start.

Dr. Matt Smith, University of Worcester

Designed for readers with little science or math background, Somerville gracefully illuminates the elements of climate science. Students will enjoy reading this book, and they will talk about it with their friends.

Professor Valerie Thomas, Georgia Institute of Technology

Climate Change Science
An Essential Reader

RICHARD C. J. SOMERVILLE
University of California, San Diego

CAMBRIDGE
UNIVERSITY PRESS

CAMBRIDGE
UNIVERSITY PRESS

Shaftesbury Road, Cambridge CB2 8EA, United Kingdom

One Liberty Plaza, 20th Floor, New York, NY 10006, USA

477 Williamstown Road, Port Melbourne, VIC 3207, Australia

314–321, 3rd Floor, Plot 3, Splendor Forum, Jasola District Centre, New Delhi – 110025, India

103 Penang Road, #05–06/07, Visioncrest Commercial, Singapore 238467

Cambridge University Press is part of Cambridge University Press & Assessment, a department of the University of Cambridge.

We share the University's mission to contribute to society through the pursuit of education, learning and research at the highest international levels of excellence.

www.cambridge.org
Information on this title: www.cambridge.org/9781009691154

DOI: 10.1017/9781009691147

© Richard C. J. Somerville 2026

This publication is in copyright. Subject to statutory exception and to the provisions of relevant collective licensing agreements, no reproduction of any part may take place without the written permission of Cambridge University Press & Assessment.

When citing this work, please include a reference to the DOI 10.1017/9781009691147

First published 2026

A catalogue record for this publication is available from the British Library

A Cataloging-in-Publication data record for this book is available from the Library of Congress

ISBN 978-1-009-69118-5 Hardback
ISBN 978-1-009-69115-4 Paperback

Cambridge University Press & Assessment has no responsibility for the persistence or accuracy of URLs for external or third-party internet websites referred to in this publication and does not guarantee that any content on such websites is, or will remain, accurate or appropriate.

For EU product safety concerns, contact us at Calle de José Abascal, 56, 1°, 28003 Madrid, Spain, or email eugpsr@cambridge.org

To Alex and Anatol

The secret is comprised in three words – Work, Finish, Publish.
—*Michael Faraday*, John Hall Gladstone

There is too much education altogether, especially in American schools. The only rational way of educating is to be an example – if one can't help it, a warning example.
—*Ideas and Opinions*, Albert Einstein

Contents

Preface		*page* ix
Acknowledgments		xvi

PART I EIGHT ARTICLES FROM THE *BULLETIN OF THE ATOMIC SCIENTISTS*

1	Reflections on the UN Climate Change Negotiations in Bali	3
2	The Passing of a Climate Prodigy	9
3	Climate Change, Irreversibility, and Urgency	13
4	Climate Change and the 2016 Election	20
5	How to Deal with Climate Change Denying Uncle Pete	23
6	Wildfires and Climate Change	29
7	Facts and Opinions about Climate Change	33
8	Weaning a House and the World from Fossil Fuels: Lessons Learned	43

PART II UNDERSTANDING AND COMMUNICATING CLIMATE CHANGE SCIENCE

9	Preparation	55
10	Stories	59
11	Language	66
12	Solutions	69

PART III SCIENTIFIC INVESTIGATIONS OF THE CLIMATE SYSTEM

13	The Greenhouse Effect	77
14	The Keeling Curve	89
15	The Temperature Record	95
16	Climate in the Future	99
17	Numerical Weather Prediction	103
18	Modeling the Climate System	108
19	Climate Feedbacks	113
20	Predictability	123
21	How Climate Models Work	135
22	The Sixth Assessment Report of the IPCC	143

PART IV THE FUTURE

23	If I Were President: A Climate Change Speech	149

Appendix: Curriculum Vitae of Richard C. J. Somerville	162
Glossary	165
Resources: Recommended Websites and Books	174
References	179
Index	181

Preface

If you enjoy reading lengthy and detailed technical reports about the physical science of climate change, I have an exciting recommendation for you. *Climate Change 2021: The Physical Science Basis* (Intergovernmental Panel on Climate Change, 2021) has been published by Cambridge University Press (CUP). It is 2,391 pages long. It fills two printed volumes. Together, they weigh almost 15 pounds, nearly 7 kilograms. If you study this one rather large report thoroughly and carefully, you may not need to read the more than 14,000 scientific publications on which it is based. To help you with some of the jargon, I will tell you that we climate scientists call this report the WGI contribution to Intergovernmental Panel on Climate Change (IPCC) AR6. That means it is the Working Group One contribution to the *Sixth Assessment Report* of the IPCC. The foreword to this impressive report begins like this: "It is unequivocal that human activities have heated our climate. Recent changes are rapid, intensifying, and unprecedented over centuries to thousands of years. With each additional increment of warming, these changes will become larger, resulting in long-lasting, irreversible implications, in particular for sea level rise."

However, if you prefer to read something about climate change that is smaller and easier to read, but still truthful, accurate, and reliable, the book you are now looking at may be helpful. This book is an anthology of readings about climate change science. I wrote them all. It is not a textbook. It is written for the general reader and is also appropriate to be used in a university course on climate change. It is intended to be especially suitable for students who have not studied university-level mathematics or science. In fact, anyone who has finished high school can understand this book.

The rationale for writing this book is that some universities are now beginning to require all undergraduate students to take an approved climate change course. There are many students who may lack strong mathematical backgrounds or may not have taken some science courses. These students may still enjoy and benefit from a good climate change course. However, they may wish to avoid courses or books that contain learning obstacles for them, such as assuming knowledge of advanced mathematics and quantitative physics and chemistry. This book is for these students.

The main material in this book is a collection of some articles and book chapters I have written about climate change. They all are written in clear, plain English for nontechnical readers. This entire book has no equations, no jargon, no complicated charts or graphs, and no math or quantitative science at all.

At the University of California, San Diego (UCSD), where I have been an atmospheric and climate science professor since 1979, there is now a new requirement for students. Beginning in 2024, every entering undergraduate UCSD student who is a candidate for a bachelor's degree must, as a requirement for graduating, take an approved UCSD course that provides substantial knowledge of climate change.

The implication is clear. I expect many other US universities will soon have similar requirements, in part because UCSD is influential. UCSD, founded in 1960, is one of the best American universities founded since World War II. However, aside from UCSD's example, other universities, including those in other countries, may also soon implement rules requiring all students to learn about climate change, simply because climate change is an important existential issue of our time.

Such a mandate, however originated, will create a need for new books explaining the science of climate change – books that must appeal to a wide variety of university students. In response to this need, I have planned this book to be useful in climate change courses for the thousands of undergraduate students who are not majoring in STEM (science, technology, engineering, and mathematics) fields. Many, perhaps most, of these non-STEM students simply cannot handle courses or books that employ a significant component of advanced mathematics or quantitative science. Of course, these students are not stupid and not inferior in any way. They simply are not enrolled in majors that require extensive coursework in science and math.

We climate scientists know that climate change is no longer a vague threat for the distant future. We are seeing climate change happen already throughout the world. Climate change is real, and it is here and now. Despite this evidence, people and governments have done far too little by way of global mitigation to reduce the threat of climate disruption.

The governments of many nations have pledged to limit climate change to no more than a global warming of 2 degrees Celsius (which is 3.6 degrees Fahrenheit) above the average temperature of the preindustrial period of the early 1800s. However, the nations of the world have not yet succeeded in making the large reductions in their emissions that would be necessary to accomplish this goal. In fact, emissions of carbon dioxide (CO_2) and other heat-trapping substances are still increasing globally, year after year. To stabilize climate at tolerable levels, these emissions must rapidly decrease and soon end completely. If people and governments fail to act decisively to reduce emissions, we run the risk of leaving an existential problem for our children and grandchildren – a problem they may well be unable to solve.

I became a meteorologist for a simple reason: weather has fascinated me ever since I was only 10 years old. More than 70 years later, now that I am semi-retired after a long career as a scientist and a professor, I am still intrigued by weather. I have spent more than 60 years studying the atmosphere and the climate system. I earned undergraduate and graduate degrees in meteorology, and I have now spent many years doing research and teaching, writing, and speaking about weather prediction and climate change.

As a child excited about weather, I could never have imagined the events that would occur during my lifetime – events that would utterly transform meteorology and climate science and would profoundly affect my career. These include technological advances. One such advance is the invention of high-speed digital electronic computers. This development began in the 1940s. These computers have revolutionized weather forecasting and have made today's complex climate models possible. Another key technological achievement is the development of artificial Earth satellites. These satellites now provide observations of the entire global atmosphere and many other components of the Earth's climate system. Both of these technological accomplishments were heavily influenced by the Cold War between the United States and the Soviet Union. This period of increased geopolitical tension began in the years immediately following the end of World War II in 1945 and lasted almost half a century, until the Soviet Union collapsed in 1991.

In 1958, the year I graduated from high school and started my university education, the US Congress established the National Aeronautics and Space Administration (NASA). The creation of NASA occurred in response to the 1957 launch of the world's first artificial Earth satellite, Sputnik 1, by the Soviet Union. When I was in my early thirties, NASA offered me my first permanent job. After earning my PhD degree and holding several temporary postdoctoral appointments, I spent my career at only three organizations. These are the Goddard Institute for Space Studies (GISS) of NASA in New York City; the National Center for Atmospheric Research (NCAR) in Boulder, Colorado; and UCSD in La Jolla, California. It is remarkable that none of these three organizations existed when I began my university education in 1958. In those days, a few visionary people were thinking about creating such organizations. However, these three places, GISS, NCAR, and UCSD, like many other wonderful centers of scientific excellence, had not yet been founded. They simply did not exist.

Being part of the scientific effort to understand and predict human-caused climate change eventually came to dominate my career. However, in 1958, when I graduated from high school and entered college, only a few scientists clearly understood that man-made increases in the amount of carbon dioxide (CO_2) in the atmosphere might cause large global climate changes.

Denying the possibility that human-caused variations in the amount of atmospheric CO_2 can produce climate changes might seem arrogant and ignorant today, but it was typical expert opinion in the 1950s. A few decades later, expert opinion had completely changed. Science relies on facts and evidence. Science advances when new facts and new evidence are discovered.

By 1978, 20 years after I began my university education, I had already become fascinated by the climate change issue. I had been impressed by some of the early research on climate change. I knew of a now-famous research article (Revelle and Suess, 1957) written by two scientists at the Scripps Institution of Oceanography, located in La Jolla, which is in the city of San Diego, California. This visionary paper contains a statement: "Human beings are now carrying out a large-scale geophysical experiment of a kind that could not have happened in the past nor be reproduced in the future. Within a few centuries we are returning to the atmosphere and oceans the concentrated organic carbon stored in sedimentary rocks over hundreds of millions of years." I also knew the conclusion that Revelle and Suess drew: "Present data on the total amount of CO_2 in the atmosphere, on the rates and mechanisms of exchange, and on possible fluctuations in terrestrial and marine organic carbon, are inadequate for accurate measurements of future changes in atmospheric CO_2. An opportunity exists during the International Geophysical Year to obtain much of the necessary information."

Revelle, who was then the Director of Scripps Institution of Oceanography, energetically seized this opportunity. He learned of a young chemist named Charles David Keeling, who was then a postdoctoral fellow at the California Institute of Technology in Pasadena. Keeling had designed and built the first scientific instrument capable of accurately measuring the amount of carbon dioxide in the atmosphere. As he would later write, "Revelle invited me to visit Scripps with the prospect of a job. I was given lunch in the back yard of the residence of his associate, Professor Hans Suess, in brilliant sunshine wafted by a gentle sea breeze" (Keeling, 1998). I am sure that Keeling was not the first or last scientist who decided to come to San Diego because of Roger Revelle's persuasive charm plus the benign climate of the region. Keeling joined Scripps Institution of Oceanography in 1956. In 1958, he began measuring the amount of atmospheric carbon dioxide accurately, an activity he would continue until his death in 2005. Charles David Keeling deserves full credit for the epochal discovery that the amount of atmospheric carbon dioxide is increasing. Keeling created a graph. This graph showed how the atmospheric carbon dioxide amount has changed with time, starting in 1958, at an observing site located on the Mauna Loa volcano in Hawaii. Today, this graph is famous. It is the most important graph in all of Earth science. It is called the Keeling curve (Figure 14.1). Important original scientific articles published by Keeling and other early pioneers in the science of climate change have been collected and are available in a fascinating book (Archer and Pierrehumbert, 2011).

I myself joined Scripps Institution of Oceanography as a professor in 1979. I soon met Charles David Keeling. We became friends and had many conversations together about climate change. Because Roger Revelle had returned to Scripps after an absence from 1963 to 1976 spent at Harvard, I also met him. We too became good friends.

In an encyclopedia article published in 2002, I wrote this:

> The Mauna Loa record of CO_2 concentration continues to the present day, and the graph of rising CO_2 concentration as a function of time over more than four decades is now known as the Keeling curve. This record, now supplemented by measurements at other sites, demonstrates conclusively that CO_2 concentrations are rising and that the primary cause is the burning of fossil fuels: coal, oil and natural gas. All discussions of the possibility of global climate change due to human activities begin with this solid empirical evidence, which no reputable scientist doubts. Charles David Keeling thus deserves credit for alerting humankind to the fact that it is changing the chemical composition of the global atmosphere. In addition, the CO_2 concentration observations have yielded a rich harvest of insights into the carbon cycle. It is important to realize that these measurements, which are now sustained by international collaboration, were originally made simply because of the dedication, perseverance and skill of one scientist. In the research community, Keeling's stubborn insistence on highly accurate and completely trustworthy CO_2 concentration measurements is legendary.

An early article (Keeling, 1970) concluded with this gloomy forecast:

> Have you noticed that practically all master plans do not project beyond the year 2000 A.D.? Our college students, however, today expect, or at least nourish the hope, to live beyond that date, and I predict that they will be the first generation to feel such strong concern for man's future that they will discover means of effective action. This action may be less pleasant and rational than the corrective measures that we promote today, but thirty years from now, if present trends are any sign, mankind's world, I judge, will be in greater immediate danger than it is today, and immediate corrective measures, if such exist, will be closer at hand. If the human

race survives into the twenty-first century with the vast population increase that now seems inevitable, the people living then, along with their other troubles, may also face the threat of climatic change brought about by an uncontrolled increase in atmospheric CO_2 from fossil fuels.

In fact, the human-caused problem of climate change was one of the most important reasons I decided to join Scripps Institution of Oceanography at UCSD. Scripps covers all the Earth sciences, and it is very strong in physical oceanography. The fact that Scripps had a world-class reputation in oceanography, and that climate change is very much an ocean-atmosphere coupled system issue, attracted me. After I joined Scripps as a professor in 1979, I changed the focus of my own research from numerical weather prediction to climate change.

As I write this in 2025, more than 60 years after Charles David Keeling began his careful measurements of atmospheric carbon dioxide amounts, it is extremely unfortunate that the world has not yet been successful in coping with the challenge of man-made climate change. The world still demands more and more energy, and fossil fuels are still the main source of that energy. As a result, the amount of carbon dioxide in the air continues to increase year after year. The resulting climate change, one that scientists had long predicted, is no longer a vague threat for a distant future. It is happening now.

I think any rational response to climate change involves first knowing what the facts and evidence are. That is the province of science. We should all learn and accept the facts and evidence, which are objective truth and should be the same for everyone. The politician and sociologist Daniel Patrick Moynihan was right when he remarked, "Everyone is entitled to his own opinion, but not to his own facts."

In my view, the most important message that climate scientists have for the public is this: *The biggest unknown about future climate is human behavior. Everything depends on what people and their governments do.*

The scientific consensus about climate change is overwhelming. Climate change is already happening, here and now. About 97% of climate experts – the scientists who are most active in carrying out and publishing research on climate change – agree that the observed recent warming is *real* and *serious* and overwhelmingly *human-caused*, and that it will become much more serious *unless* humanity makes big changes very soon in how it generates energy.

My own career has evolved too, sometimes in ways that surprised me. Most of the research I have worked on recently differs greatly from the topics that dominated my professional life decades ago. There have been many changes in atmospheric and climate science since I was a student. The technology and knowledge available today allow us to make real progress now on scientific problems that would have seemed unsolvable when I was much younger. Also, much of life is unpredictable. Seemingly small choices may have large consequences, and luck always plays a role. I have found deep professional and personal satisfaction in being a professor and a climate scientist for many years.

I have chosen two epigraphs to appear in the front of this book. One of them is the famous advice to scientists attributed to Michael Faraday, the brilliant nineteenth-century physicist and chemist (Gladstone, 1873): "Work, finish, publish." I have a few words to say about Faraday and the advice he gave. I think the other epigraph (Einstein, 1954) is crystal clear and needs no explanation from me.

Advice can often be easy to state, but for ordinary mortals, Faraday's advice would obviously not enable them to achieve results as brilliant as Faraday's in physics and chemistry. Faraday may not be very well known to the general public today, but he is still an inspiring hero to many scientists. In his study, Albert Einstein kept portraits of three great scientists whom he revered: Isaac Newton, James Clerk Maxwell, and Michael Faraday. Two of these three, Newton and Maxwell, like Einstein himself, were gifted theoreticians with powerful mathematical abilities. Faraday, however, was a different kind of scientist. Faraday had educated himself by reading widely. He had little formal education, and his mathematics was limited to simple algebra, which students today can learn in high school. Faraday is a wonderful role model for anybody with scientific aspirations who finds learning mathematics to be a formidable obstacle.

Despite this educational handicap, Michael Faraday (1791–1867) became one of the greatest experimental physicists and chemists of all time. His physical intuition was extraordinary, and his experimental abilities were unsurpassed. Faraday's epochal discoveries in electromagnetism and related areas led to Maxwell's classical mathematical theory of electromagnetic radiation, optics, and electric circuits, as Maxwell himself often acknowledged. Yet, Faraday was a modest and deeply religious man, and he was surely sincere in giving his simple but famous advice: "Work, finish, publish." I have chosen his words as an epigraph for this book to emphasize how I think climate scientists can best contribute to helping the world meet the great challenge of climate change. I think we should all do the best science we are capable of doing, and those of us who can write and speak about it to the public should also do that often to help inform the world. Everyone who cares about climate change can communicate it. If enough people are well informed about climate change, and if they care enough about the future that their children will inherit, then the world may act with sufficient energy and decisiveness. *Everything depends on what people and their governments do.*

If you have decided to read this book, thank you. I welcome you to a fascinating subject.

Acknowledgments

I owe a great debt to the many graduate students, postdoctoral fellows, research scientists, and other friends and colleagues with whom I have worked.

I especially thank Carolyn Baxter, my superb secretary who managed my office throughout my 40-year career at Scripps Institution of Oceanography, UCSD. Our long partnership has been consistently warm and unfailingly pleasant. With wisdom and tact, Carolyn has deftly handled budgets, conferences, correspondence, manuscripts, progress reports, research proposals, visitors, and voluminous red tape. Thanks to her, I and everyone else in my group could concentrate on science.

I am grateful to Dr. Michelle Niemann for editing an early manuscript of this book, as well as some of my earlier writing. All writers, whether they know it or not, need an excellent editor. I have been fortunate to find one.

Several friends and colleagues looked at various drafts of this book and made many helpful suggestions. I thank Robert Chervin, Michael Edesses, Kerry Emanuel, Danielle Lafarga, Anamaria Navarrete, and Gerald North for being especially helpful.

Preparation of the manuscript and its editing have been supported in part by a Spilhaus Ambassador Award Grant from the AGU.

For granting me permission to include several examples of material that I had previously published elsewhere in this book, I gratefully acknowledge the American Institute of Physics, the AMS, the *Bulletin of the Atomic Scientists*, the Regents of the University of California, and the University of California Press.

I am grateful to CUP for publishing this book, and especially to Dr. Matt Lloyd, my CUP editor. He guided me skillfully throughout the long and complicated process of turning an idea into a book. I also thank Maya Zakrzewska-Pim of CUP for being especially helpful. Comments on an early version of the manuscript, by four anonymous reviewers selected by Dr. Lloyd, led to significant improvements in this book.

Sylvia Francisca Bal Somerville, whose husband I have been for 60 years, continues to love, support, and tolerate me, for which I am deeply grateful. Sylvia is Dutch. She was born in Amsterdam at the start of World War II. During the war, she and her family suffered from the cruel and brutal occupation of the Netherlands by Nazi Germany. Sylvia vividly remembers the fierce fighting in the battle of Nijmegen late in the war. Her entire family miraculously came through the war alive. Many Dutch civilians did not. There have been several times in our life together when various difficulties in my career seemed unsolvable. Sylvia's perspective at these times has always comforted and reassured me. She would simply say, "I survived Hitler. We will survive this." She has always been right.

PART I

Eight Articles from the *Bulletin of the Atomic Scientists*

Introduction to Part I

During World War II, the Manhattan Project was a secret military research effort in the United States. This project developed the first atomic bomb. A decision to use this weapon against Japan was made by President Truman. Atomic bombs were dropped on the cities of Hiroshima and Nagasaki in August 1945. Although exact numbers of deaths are unknown, it has been estimated that approximately 70,000 to 135,000 people died in Hiroshima and 60,000 to 80,000 people died in Nagasaki. These two attacks and other heavy losses caused Japan to surrender unconditionally within a few days. Because Germany had already been defeated earlier in 1945, the Japanese surrender effectively ended the war.

The *Bulletin of the Atomic Scientists* was founded late in 1945. Albert Einstein was involved, as were many scientists who had worked on the development of the atomic bomb. I published several articles in the *Bulletin of the Atomic Scientists* and have reprinted eight of them here. They were published between 2008 and 2022 and are reproduced here in chronological order. I am grateful to the *Bulletin of the Atomic Scientists* for granting me permission to use these articles in this book. I have slightly modified some of the articles in order to update the science, improve clarity, and reduce redundancy.

1
Reflections on the UN Climate Change Negotiations in Bali

In recent years, I have been to many of the yearly United Nations Conference of the Parties (COP) meetings. In this case, the document that nearly all countries are parties to is called the UN Framework Convention on Climate Change (UNFCCC). This document was signed at the Earth Summit, a famous international event held in Rio de Janeiro, Brazil, in 1992. At the time, the Earth Summit was the largest gathering of world leaders ever held. Representatives of 178 nations participated in the Earth Summit, and 117 heads of state attended. The objective of the UNFCCC is "stabilization of greenhouse gas concentrations in the atmosphere at a level that would prevent dangerous anthropogenic [i.e., human-caused] interference with the climate system." COP meetings have been held in most years since the Earth Summit. The first COP meeting, known as COP 1, was held in Berlin in 1995. There have been almost 30 COP meetings since then. Each COP meeting attracts thousands of people and lasts about 12 days. These COP meetings have provided opportunities for countries to discuss topics such as the reduction of global emissions of carbon dioxide and other heat-trapping substances.

COP 13 was held in Bali, Indonesia, in December 2007. I was there. Two months earlier, the world had learned that the 2007 Nobel Peace Prize was to be awarded equally between Al Gore, the American politician, and the Intergovernmental Panel on Climate Change (IPCC). I think it is safe to say that in 2007, almost every American knew something about Al Gore, who had been the US Vice President from 1993 to 2001 and was the unsuccessful Democratic party candidate for President in the 2000 election. Gore had been a student at Harvard from 1965 to 1969. He had wide interests but was said not to do well in science classes, and he did not take mathematics courses. However, in his senior year at Harvard, Gore attended a course taught by Roger Revelle, an oceanographer who had been Director of Scripps Institution of Oceanography and who had been the leader in the founding of the University of California,

San Diego. Al Gore later spoke often of his high regard for Roger Revelle and always credited Revelle as being the initial inspiration for Gore's intense interest in climate change and other environmental issues. The 2007 Nobel Prize announcement said of Gore, "He is probably the single individual who has done most to create greater worldwide understanding of the measures that need to be adopted."

Gore was indeed well known in 2007. However, I am confident that in 2007 relatively few Americans knew anything about the IPCC, the organization that shared the 2007 Nobel Peace Prize with Al Gore. As for me, in 2007 I had just finished three years of working almost full time for the IPCC, and I was thrilled by the Nobel Prize award. I had been Coordinating Lead Author for the 2007 IPCC *Fourth Assessment Report*. The chapter of this report that I had been a coleader of was called *Historical Overview of Climate Change*. I was also one of the signers of the *2007 Bali Climate Declaration by Scientists*. In this chapter, I tell the story of the IPCC *Fourth Assessment Report* and the COP 13 meeting in Bali.

Reflections on the UN Climate Change Negotiations in Bali (2008)

By Richard C. J. Somerville, January 7, 2008

(This article from the *Bulletin of the Atomic Scientists* is reproduced by permission)

The IPCC, long regarded as the single most trustworthy source of information on climate science, states unequivocally that the Earth's climate is warming rapidly and that we are now more than 90% certain that human activities have caused most of the observed warming in recent decades. The research behind these findings, first published in the IPCC's landmark 2007 *Fourth Assessment Report* (known as AR4), is rock-solid science. In recognition of the importance and excellence of its work in bringing climate science to the attention of policy-makers and the public, the IPCC shared the 2007 Nobel Peace Prize. The prize was awarded jointly to the IPCC and to Al Gore, the US Vice President from 1993 to 2001, "for their efforts to build up and disseminate greater knowledge about man-made climate change, and to lay the foundations for the measures that are needed to counteract such change."

Everybody at the UN Framework Convention on Climate Change (UNFCCC) negotiations, which took place in December 2007 in Bali, Indonesia, was aware of the IPCC's latest conclusions. These negotiations are an annual event, and the Bali meeting attracted more than 10,000 people, including a large press corps and many stakeholders from the energy industry and environmental interest groups. I attended COP 13 in Bali representing a group of 200 climate science experts. We wanted to inject some quantitative scientific substance into the negotiations.

Climate science is now able to provide significant predictive power, meaning the science summarized in the 2007 IPCC AR4 report can link specific concentrations of greenhouse gases with the associated climatic consequences. How much sea level will rise or how much temperatures will change *does* depend on how much humanity allows greenhouse gas amounts to increase. Yet, what is said publicly by the negotiators almost never refers to the science in this way. Thus, in 2008 the term "dangerous" in the UNFCCC objective of preventing dangerous anthropogenic interference with the climate system remained quantitatively undefined.

As a veteran observer of these difficult negotiations, I was reminded of the dark days of an earlier environmental crisis, the human-caused depletion of the stratospheric ozone layer, dramatized by the creation of an ozone hole over

Antarctica. In the mid-1980s, frustrated by the lack of progress in reaching an agreement to end the manufacture of the chemicals that caused the problem, a prominent US atmospheric scientist, F. Sherwood Rowland, who would later share the Nobel Prize in chemistry, told a journalist, "What's the use of having developed a science well enough to make predictions, if in the end, all we're willing to do is stand around and wait for them to come true!"

Today, I feel exactly the same way about the procrastination and posturing that too many governments have substituted for meaningful action to limit global warming. The statements attributed to the United States, especially by Paula Dobriansky, who headed the US delegation, and James Connaughton of the White House Council on Environmental Quality, have been widely reported, and have clearly caused offense to many countries. But I do not want to single out the United States. China was notably unwilling to accept any restrictions on its greenhouse gas emissions, and several other countries followed or welcomed the US lead. The struggle to have major countries agree to specific targets and timetables has been unsuccessful so far.

Meanwhile, the rate at which humankind is emitting greenhouse gases into the atmosphere continues to increase every year. Only a sharp reversal of this trend offers hope of stabilizing the amount of these gases in the atmosphere and limiting climate change.

The international group that I represented in Bali went to publicize "The 2007 Bali Climate Declaration by Scientists." Signatories of this document include many leaders of the climate science research community whose excellence and expertise are beyond doubt. They include Nobel laureate Paul Crutzen; National Academy of Sciences members Robert Dickinson, Kerry Emanuel, Stephen Schneider, and Carl Wunsch; quite a few lead authors and coordinating lead authors of the IPCC *Fourth Assessment Report*; and several heads of major research institutions. But we each signed the declaration representing only ourselves, not the IPCC, our employers, or any other organization.

"The 2007 Bali Climate Declaration by Scientists" had the immediate goal of asking the negotiators to reach an agreement limiting global warming to 2 degrees Celsius (3.6 degrees Fahrenheit) above the preindustrial temperature. Global warming is simply a symptom or measure of the magnitude of climate change, and 2 degrees Celsius is widely thought to be a low enough reasonable consensus estimate to prevent dangerous climate change. In fact, the European Union and other countries have already formally adopted this limit.

What actions are needed to ensure that the 2-degree Celsius target is not exceeded? This is where climate science has a key role to play. While we cannot know safe greenhouse gas concentrations or emission limits exactly – just as medical science cannot specify a precise value as a safe limit on, say, cholesterol – the science summarized in the 2007 IPCC report does provide excellent quantitative guidance.

The IPCC, which is mandated to be policy-neutral, cannot make policy recommendations. However, once a policymaker accepts 2 degrees Celsius above preindustrial levels as a target upper bound on allowable warming, then the IPCC report describes approximately what the limit must be for greenhouse gas concentration, in what year carbon dioxide emissions must peak before declining, and what percentage reduction in emissions is needed. (The information is succinctly summarized in table SPM.5 in the Working Group III *Summary for Policymakers* and is repeated as table 5.1 in the *Fourth Assessment Synthesis Report*.)

Taking a 2-degree-Celsius limit on global warming as the goal, the Bali declaration states,

> Based on current scientific understanding, this requires that global greenhouse gas emissions need to be reduced by at least 50 percent below their 1990 levels by the year 2050. In the long run, greenhouse gas concentrations need to be stabilized at a level well below 450 parts per million. In order to stay below 2 degrees Celsius, global emissions must peak and decline in the next 10 to 15 years, so there is no time to lose. As scientists, we urge the negotiators to reach an agreement that takes these targets as a minimum requirement for a fair and effective global climate agreement.

The outcome of the Bali meeting was deeply disappointing, but not surprising. The science cannot make a subjective judgment on how much warming is acceptable or say in what ways society should stabilize greenhouse gas concentrations. Those are political judgments. In the end, after much bitterness, the negotiators reached an agreement to continue negotiating over the next two years, but they did not come close to agreeing on the substance of such an agreement. It is true that many international agreements took quite a few years to materialize. What is clear scientifically, however, is that time is running out. If an agreement is not possible in the next two, three, or four years, it may be too late to prevent serious climatic consequences.

At subsequent climate negotiations, the important question will be whether governments are willing to implement and enforce an effective agreement to halt the rapid rise in atmospheric carbon dioxide and other pollutants that cause global warming. My personal view is that governments, especially democratically elected ones, will respond to the will of their people. And what will it take to make large numbers of people increase the priority that they now give to this issue? Perhaps it will take some sudden, shocking, and unambiguous climate event such as the destabilization of a large part of the Greenland ice sheet and a sharp increase in sea level. The ozone hole would be a parallel case historically. It would be a pity if we needed to wait for that, which is like waiting to have a heart attack rather than heeding a physician's warnings about cholesterol and weight. In climate science as in medical science, prevention is better than a cure, but not everybody is wise enough to act early.

Governments and businesses worldwide, and ultimately, humankind as a whole, will determine what actions will be taken. Climate science, however, is able to provide highly useful input to this policymaking process. Making wise climate policy includes taking sound climate science into account. The future depends on it.

2
The Passing of a Climate Prodigy

Young students, like all learners, arrive at the classroom with prior knowledge of the subject, which can be accurate or inaccurate and more or less relevant to the subject. Constructivist learning theory has taught us that the mind of the student is not an empty vessel to be filled, but that understanding the initial contents of the student's mind is critical to the educational process. A young student in particular may well be unaware that the science of climate change has ancient roots. A few scientists in the 1800s did pioneering research.

The threat of climate change was indeed becoming apparent to some climate scientists by the 1970s. However, the scientific community had not yet brought the details of this threat to the attention of the world. Stephen H. Schneider was among the very first climate scientists to dedicate himself to this task and to become proficient at it. I knew Steve Schneider well, as a good friend and scientific colleague, from his days as a graduate student in the 1960s until his death at age 65 in 2010. He encouraged and inspired me and many other climate scientists to learn how to communicate climate science more effectively, how to collaborate well with persons working in the mass media, how to interact usefully with those in the political sphere, and how in general to bridge the gap between the scientific research effort and the wider world.

I want the readers of this book to understand that this effort at climate change science communication is not something novel and recent, but it has been actively progressing via many of us climate scientists for at least half a century. I also want to emphasize that climate change science communication needs to continue, because the world has not yet acted forcefully and effectively enough to limit climate change to an amount that nations of the world have agreed on. As I will repeat more than once in this book, "Everything depends on what people and their government do."

The Passing of a Climate Prodigy

By Richard C. J. Somerville, July 23, 2010

(This article from the *Bulletin of the Atomic Scientists* is reproduced by permission)

A towering figure in climate science, Stephen H. Schneider, aged 65, died after suffering a pulmonary embolism on July 19, 2010, while flying to London from a conference in Stockholm. The loss of Schneider, a professor at Stanford University, deprives the world of both an outstanding researcher and a gifted science communicator. To his colleagues in climate science, Steve, as everybody called him, has long been known as a scientific pioneer and a role model. For four decades, he tirelessly carried out research, explained climate science to the wider world, and advocated rational responses to the threat of human-caused climate change.

Schneider, born in New York City, was first attracted to the infant science of climate change while a graduate student at Columbia University in the late 1960s. There he studied mechanical engineering, fluid dynamics, and plasma physics. Climatology in the 1960s was still largely a scientific backwater devoted to compiling weather statistics. Schneider was at the center of what was to become a genuine scientific revolution. In 1971, while a postdoctoral fellow at the NASA Goddard Institute for Space Studies (GISS) in New York, he published one of the first attempts to model the effects of carbon dioxide and atmospheric aerosols on climate. In 1972, Schneider left GISS for the National Center for Atmospheric Research (NCAR) in Boulder, Colorado, where he stayed for more than 20 years before moving to Stanford. At NCAR, he was among the first to recognize and analyze the large potential influence of cloud feedbacks on climate change.

Schneider quickly demonstrated talents and traits that would endure for life. He was an intense, driven, high-energy scientist, who, well before most, realized that climate science was inherently interdisciplinary. He was always highly collaborative, working productively with an extremely wide range of scientists. He was drawn toward the world of science policy and the broad implications of climate change for society. He was unfailingly generous and altruistic in supporting and counseling other scientists. Many leaders in the climate science community today credit him with influencing and aiding their careers. His integrity – both personal and scientific – was absolute. He had a prodigious work ethic and published prolifically throughout his career. He was courageous when facing adversity and invariably courteous to those who disagreed with him.

It must be said that Schneider also had an ego, talked nonstop, and loved being on television. If these are character flaws, then we surely need more flawed characters just like him. Journalists quickly learned that Schneider was the scientist to seek out whenever a complex climate science story needed to be told. They would immediately get an articulate answer from him to any question. It would be free of jargon, scientifically accurate, direct, and to the point. Schneider invented the term "mediarology" and later built a section of his Stanford website to explain how scientists could more effectively interact with journalists.

At NCAR, Schneider was a key participant in forming a new climate science program in 1972, while he was still a postdoctoral fellow. In 1975, at age 30, he boldly founded a new interdisciplinary journal, *Climatic Change*. He had edited this journal for the rest of his life. In recent years, he quietly subsidized the salaries of its small staff out of his own pocket. Schneider also published several books popularizing climate science for a general audience, beginning with *The Genesis Strategy* in 1976. This book predicted that greenhouse gases would cause a "demonstrable" climate change by the end of the century. Characteristically, Schneider was ahead of his time.

As climate change became a subject of great scientific interest as well as public concern, the obvious excellence of Schneider's research and his ability to explain science to the wider world solidified his outstanding international reputation. The creation of the Intergovernmental Panel on Climate Change (IPCC) in 1988 marked global recognition of the need to bring the best scientific information to the attention of policymakers. Schneider's wide-ranging knowledge, and his ability to move easily between the worlds of physical science, social science, policymakers, and the press, made him an indispensable asset to IPCC. He was a lead author for all four of the major IPCC assessment reports to date, as well as for two synthesis reports. He consulted for every US administration since that of Richard Nixon and frequently testified before Congress.

In 2001, Schneider was diagnosed with mantle cell lymphoma, an especially aggressive life-threatening cancer. This is a rare type of non-Hodgkin's lymphoma for which there was no standard cure and for which very few clinical trials data were available. Schneider, typically rational, found parallels with problems in climate science and decided to partner actively with his medical team in designing his treatment path. He later said that his doctors explained oncology to him while he explained Bayesian statistical inference theory to them. With strong support, especially from his wife, Terry Root, herself an eminent scientist and frequent research collaborator with Schneider, the treatment was successful. For him, the long ordeal was a teachable moment, and he produced an inspiring book about it, *The Patient from Hell*. Despite the painful and exhausting treatment, involving radiation, chemotherapy, and bone marrow transplants, he continued to work from the hospital.

Schneider was a member of the *Bulletin*'s Science and Security Board. Recognized by many honors, he received a MacArthur Fellowship "genius award" in 1992 and was elected to the National Academy of Sciences in 2002. His last book, *Science as a Contact Sport*, a scientific autobiography, appeared in 2009. A recounting of his four decades of experience in both climate science and climate policy, it is the single best source for those who want to know more about Schneider as a scientist and a public intellectual. For those of us who were fortunate enough to have been Schneider's colleagues and friends: Although his eloquent voice has now been silenced, his powerful influence on us all is indelible, and the example of the life he led will continue to be an inspiration.

3
Climate Change, Irreversibility, and Urgency

If I were asked to identify a single critically important aspect of modern climate change science that many people seem unable to grasp, it would be the urgency of acting to limit the climate change. This urgency has nothing to do with politics or economics. Instead, it arises directly from the physics and chemistry of the climate system. Carbon dioxide, once emitted, remains in the atmosphere for a very long time. A significant portion of it can persist in the atmosphere for centuries and even millennia until natural processes remove it. Thus, it will be there essentially forever, if we think in terms of the implications on human time scales. No practical and sufficiently inexpensive method has yet been found to remove large amount of carbon dioxide from the atmosphere. The only known way to prevent atmospheric carbon dioxide level from increasing further is simply to cease emitting carbon dioxide into the atmosphere.

Humanity thus must find a way to reduce global emission amounts drastically and to do it soon, once a target for allowable amounts of climate change is agreed to by the nations of the world. The Paris Agreement of 2015 produced such a target. I describe the significance of the Paris Agreement elsewhere, especially in Chapter 23 in the discussion of Figure 23.1. This is a key example of why the science is directly relevant to policy, while not being at all prescriptive of policy. The science does not say, for example, whether nuclear electricity is or is not a good idea, or whether cap-and-trade systems are advisable, but the science does make it clear that waiting a long time to reduce emissions is unwise, because the resulting large climate change cannot be reversed on human time scales by any method that has yet been developed.

Climate Change, Irreversibility, and Urgency

By Richard C. J. Somerville, August 13, 2012

(This article from the *Bulletin of the Atomic Scientists* is reproduced by permission)

Climate scientists like to think of themselves as wise planetary physicians, explaining to the world what they have learned about climate and advising humanity on how to cope with the challenge of climate change. This metaphor can also appear attractive to policymakers and the public. Consider the appealing similarities between deciding what you should do about your weight and what the world should do about global warming. You can ask your doctor's opinion, but it is you who will determine your target weight. You can also ask your physician to recommend actions to reach that target. You can then experiment with diet and exercise, evaluate the results, and make changes. Throughout the process, you are in charge, and the physician's role is simply to advise.

There are obvious parallels with the climate change issue. Instead of body weight, the global average temperature at the surface of the Earth is the main metric. In place of your physician, the world has the combined expertise of the global community of climate scientists, as assessed by national academies of science and the Intergovernmental Panel on Climate Change. Deciding what climate change policies to adopt is up to the governments of all nations. Their first task is to decide how much global warming they regard as tolerable. Governments have scientific input, but they will also consider risk tolerance, national priorities, economic impacts, and political considerations.

Suppose the world's governments could eventually manage to agree on a specific aspirational goal, such as limiting global warming to 2 degrees Celsius (3.6 degrees Fahrenheit) above the average temperature of the mid-1800s, a time before human activities had begun to affect the global climate significantly. In fact, the Paris Agreement of 2015 does set this target. Governments could next ask climate scientists what reductions in the emission rates of carbon dioxide and other heat-trapping gases would be needed to meet the 2-degree Celsius target. Then governments could negotiate with one another, aiming for a binding international agreement, with specific and enforceable emission-reduction levels and timetables for all countries. Science will thus have informed this process, but governments will have made all the policy decisions.

Can this approach succeed? Some environmental challenges have followed this script faithfully. One example is the apparently successful attempt to limit stratospheric ozone loss caused by manmade chemicals such as chlorofluorocarbons. Scientific research established the cause of the ozone loss. Then the political process – with substantial scientific input – quickly led to the Montreal Protocol and subsequent international agreements to halt the manufacture of ozone-depleting chemicals. This paradigm is often cited as a role model for limiting climate change to moderate levels.

The stratospheric ozone story unfolded quickly. The key scientific paper predicting ozone loss due to chlorofluorocarbons appeared in 1974, the discovery of severe Antarctic ozone depletion was published in 1985, and the Montreal Protocol entered into force in 1989. The international community, however, has already taken much longer, both scientifically and politically, to come to grips with the climate change issue, and the doctor's advice for dealing with the malady cannot be ignored much longer. A failure to reduce carbon dioxide emissions significantly and soon – soon being within a decade or so – will have large adverse effects that are, because of the physics and biogeochemistry of the climate system, essentially irreversible on human time scales. This is not an ideological or political assertion but an unavoidable consequence of a simple reality: Once excess carbon dioxide is added to the Earth's atmosphere, much of it will remain there for centuries or longer before natural processes remove it.

In hindsight, it is easy to see that mankind had clear early warning about the need to rapidly reduce emissions of heat-trapping gases in order to avoid climate disruption. In the 1970s, two Swiss scientists carried out research that anticipated the challenge now confronting policymakers. Hans Oeschger and his doctoral student, Ulrich Siegenthaler, were pioneers in the development of Earth system models. Their paper (Siegenthaler and Oeschger, 1978), based on simple models and limited data, concluded that "For a prescribed maximum increase of 50 percent above the preindustrial carbon dioxide level, the production [of carbon dioxide emissions] could grow by about 50 percent until the beginning of the next century, but should then decrease rapidly." The most recent research, with vastly improved data and models incorporating a far more complete understanding of the climate system, has refined the quantitative conclusions and confirmed that essential result. Unfortunately, however, global emissions of carbon dioxide have not decreased. Emissions have increased instead. The resulting climate disruption has begun, and it continues.

The political process, by contrast, has been extremely slow. The Earth Summit held in Rio de Janeiro in June 1992 was attended by more than 100 heads of nations and was widely praised as a landmark in environmental statesmanship. This conference created the United Nations Framework

Convention on Climate Change, a treaty signed by nearly every country on the Earth. The objective of the treaty is to stabilize the amount of greenhouse gases in the atmosphere "at a level that would prevent dangerous anthropogenic interference with the climate system." Deciding what that "dangerous" level should be, however, is a task that the Rio negotiators left for the future.

Some 20 years later, most governments have now converged on a working definition of "dangerous." These governments have agreed that limiting the increase in the average surface temperature of the Earth to 2 degrees Celsius above preindustrial levels would be a tolerable amount of global warming. Of course, this is a subjective decision, and some governments have argued for a lower and safer amount of warming. Officially, the European Union has accepted the 2-degree Celsius target as a concrete policy objective. Less formally, the Copenhagen Accord, endorsing the 2-degree goal, has been agreed to by some 141 countries, which together account for the great majority of the world's emissions of heat-trapping gases. This agreement, reached at December 2009 negotiations in Copenhagen, is not legally binding, however, and sets no concrete targets for emissions reductions.

And it must be said that very little significant progress has occurred toward actually making the large cuts in global emissions of heat-trapping gases that would be needed to meet the 2-degree goal. In fact, the annual amount of global carbon dioxide emissions from human activities is now (in 2012) about 50% larger than that in 1992. These emissions increased the amount, or concentration, of carbon dioxide in the atmosphere by 10% during this 20-year period. This amount is now about 40% higher than it was in the mid-1800s. It is also an observed fact that the world has already warmed by about 0.8 degrees Celsius relative to the mid-1800s. That is nearly half the magnitude of warming judged to be tolerable by the Copenhagen Accord. Meanwhile, research and recent observations of the changing climate have shown that the effects of a 2-degree Celsius warming may be far more disruptive than had earlier been thought. For example, in the range of 1–2 degrees Celsius warming, the area burned by wildfire in parts of western North America is expected to increase by a factor of 2–4 *for each degree* of warming. Other climate consequences in this range of seemingly moderate warming include large changes in precipitation, increase in extreme weather events, loss of food production, rise in sea level, reduction in stream flow in many river basins, and loss of Arctic sea ice.

The outlook is grave. Recent research indicates that the climate in coming decades and centuries will be largely determined by human activities. The natural factors that have led to large climate changes in the distant past, such as changes in solar luminosity and in the Earth's orbit, obviously still exist. These natural factors are simply too weak and too slow, however, to be significant over time scales of decades to centuries. On these time scales, human activities will dominate over natural factors. In particular, carbon dioxide is by far the most important heat-trapping gas produced by human activities.

However, in addition to carbon dioxide, several other substances that are emitted because of human activities can also increase the greenhouse effect and cause warming. These include the gases methane, nitrous oxide, and ozone. Additionally, some small particles in the atmosphere, called aerosols, including black carbon, a component of soot, can also absorb radiation and thus add to the greenhouse effect. Other atmospheric aerosols can have a cooling effect. Aerosols can also affect clouds, and clouds can both warm and cool. Water vapor, the gaseous form of water, is an important greenhouse gas. Its atmospheric amount varies greatly and is mainly controlled by temperature. Thus, adding carbon dioxide warms the atmosphere, which can cause an increase in water vapor amount, thus further warming the atmosphere. This is an example of a feedback, a technical term for a sequence of events in which a cause produces a result, and the result then in turn influences its cause.

In brief, the connection between atmospheric carbon dioxide amount and climate, while extremely complex in its details, can be summarized as follows: A substantial fraction of the carbon dioxide emitted into the atmosphere by human activities remains in the atmosphere for centuries or longer. The winds distribute carbon dioxide around the globe, resulting in a uniform concentration for climate modeling. Even in the inconceivable event that emissions were to suddenly stop completely, the climate changes caused by carbon dioxide would persist for a very long period of time. The cumulative amount of carbon dioxide emitted by human activities since the industrial revolution of the 1800s will largely determine the magnitude of the resulting climate change. For any specific target limit to global warming, such as 2 degrees Celsius, there is thus an allowed amount of cumulative carbon dioxide emissions that must not be exceeded.

The quantitative implications of this scientific understanding are sobering. They confirm the insights of Siegenthaler and Oeschger (1978). If global emissions of heat-trapping gases were to continue at current rates for the next 20 years, until about 2032, and then immediately cease entirely, the cumulative emissions will have exceeded the best estimate of the allowed amount, so the likelihood of meeting the 2-degree Celsius target would be low. Instead, in order to limit warming to 2 degrees Celsius, global emissions need to peak very soon and then must decline rapidly. To stabilize climate, an almost completely decarbonized global world – with near-zero emissions of carbon dioxide and other long-lived greenhouse gases – needs to be achieved well within this century. More specifically, the global average annual per capita emissions will have to shrink to well below one metric ton of carbon dioxide by 2050. We are not making progress toward this goal. The data show that between 1960 and 1970, this measure increased from about 3 to 4 metric tons. By 2005 it had increased to about 4.5 metric tons. Between 2005 and 2023, it fluctuated between 4.5 and 4.9 metric tons. This number is emissions per person, and we must keep in mind that global population also increased substantially during this period, from about 3 billion in 1960 to about 8 billion in 2023. The total amount of global emissions per year thus continues to grow, not shrink.

The climate system dictates a timescale for action. Decisions made by governments early in this century will have long-term consequences for the climate. Failure to take meaningful actions to reduce global emissions is a particularly serious decision. By continuing on its current path of increasing global carbon dioxide emissions, humanity is committing future generations to a strongly altered climate. Even beyond the current century, there are major implications for longer-term climate change. Higher temperatures and accompanying changes in climate caused by carbon dioxide emissions from human activity will be largely irreversible on human time scales. The effects are long lasting. Atmospheric temperatures are not expected to begin to decrease substantially for many centuries to millennia, even after human-induced greenhouse gas emissions stop completely.

To meet the 2-degree Celsius target, the climate system itself thus imposes a timescale on when emissions need to peak and then begin to decline rapidly. Scientific understanding of this timescale shows that mitigation of climate change is urgent and cannot wait. A window of opportunity to decrease emissions has been open for several decades but will soon close and will remain closed. This urgency is thus not ideological or political but rather is due to the long atmospheric residence time of atmospheric carbon dioxide and the resulting near irreversibility of climate change for centuries or longer. This is a critical shortcoming of the medical metaphor that likens the climate change problem to the task of an individual trying to reach a target body weight. If your current weight is higher than your target, medical science can advise you on how to lose weight. If the amount of carbon dioxide in the atmosphere becomes higher than the amount compatible with the desired limit on global warming, however, the warming will exceed the target. There is no proven technology to remove a large amount of carbon dioxide from the atmosphere economically.

3 Climate Change, Irreversibility, and Urgency

In short, the urgency of acting to mitigate climate change is directly linked to the physics and biogeochemistry of the climate system. An enduring failure to achieve meaningful science-based international agreements to rapidly reduce emissions of heat-trapping gases will inevitably have serious consequences for the degree of climate change that the Earth will undergo. If the world as a whole continues to procrastinate throughout the current decade, allowing emissions to continue to increase year after year, then it will almost certainly have lost the opportunity to limit warming to 2 degrees Celsius. Subsequently, our children and their descendants, and ultimately all living things, will have to face the consequences of more severe climate disruption.

4
Climate Change and the 2016 Election

It is a sad fact that climate change has become a partisan issue, at least in the United States. This was not always the case. Donald Trump, the 45th and 47th president of the United States, has for many years repeatedly expressed his opinion that human-caused climate change is not a dangerous threat and that anyone who accepts the fundamental results of mainstream climate change science is badly mistaken.

Catherine Gautier, the coauthor of this article, is an atmospheric scientist who is Distinguished Professor Emerita at the University of California, Santa Barbara. Dr. Gautier and I were both accredited as journalists attending COP 21, held in Paris, France, in December 2015. We published this article in March 2016, shortly after the conclusion of COP 21.

At the COP 21 United Nations climate conference, almost all the nations of the world adopted the Paris Agreement, a legally binding international treaty aimed at limiting climate change. The temperature goal of the Paris Agreement is to limit the increase of the global average surface temperature to well below 2 degrees Celsius (or 3.6 degrees Fahrenheit) above the pre-industrial level.

The United States withdrew from the Paris Agreement in 2020, rejoined the agreement in 2021, and again announced its withdrawal from the agreement in 2025.

Climate Change and the 2016 Election

By Richard C. J. Somerville and Catherine Gautier,
March 13, 2016

(This article from the *Bulletin of the Atomic Scientists* is reproduced by permission)

As this is written, all four remaining candidates in the race for the 2016 Republican presidential nomination vehemently reject the fundamental findings of modern climate science. These findings are simple to state:

> The Earth's climate is now unequivocally warming. Many chains of evidence demonstrate the warming, including increasing atmospheric and ocean temperatures, rising sea levels, shrinking glaciers and ice sheets, and changing precipitation patterns. The main cause of the warming is human activities, especially burning fossil fuels, which increases the amount of heat-trapping gases in the atmosphere. The effects of man-made climate change are already being felt, and they are mainly harmful. The consequences of climate change will become much more severe in the future, unless global actions are taken soon to drastically reduce the amount of heat-trapping gases emitted into the atmosphere.

These conclusions are the results of decades of research by the international scientific community. They have been endorsed by the Intergovernmental Panel on Climate Change, and by the National Academy of Sciences and leading scientific professional societies in the US and other countries. The great majority of mainstream climate scientists find these results persuasive.

Nevertheless, Ted Cruz heaps scorn on what he has called "a pseudo-scientific theory." He has dismissed it as, "not science, it is a religion." John Kasich says, "I do not believe that humans are the primary cause of climate change." Marco Rubio agrees, stating, "I do not believe that human activity is causing these dramatic changes to our climate the way these scientists are portraying it." Donald Trump speaks of a "global warming hoax," calling it, "created by and for the Chinese."

These public figures reject mainstream climate science because they view it through a lens that incorporates their firmly held values and convictions. They have a high regard for American capitalism and private industry, or the free enterprise system, and a low regard for taxes and regulation, which they regard as government interference. In rejecting mainstream science, they are expressing their opposition to policies that governments might implement, if the science were accepted.

In the United States, aspiring Republican politicians may also feel the pressure to conform to a litmus test. In order to obtain political and financial support, especially from sources allied with the fossil fuel industry, they may conclude that they must attack mainstream climate science and insist that man-made climate change is not a problem.

However, Mother Nature, or the physical climate system, is not concerned with anybody's values or convictions or political litmus tests. Mother Nature is concerned with natural laws. Heat-trapping gases in the atmosphere do trap heat. That leads to warming. After every politician has expressed an opinion, Mother Nature bats last.

Yet the Republican Presidential candidates have gone to a great deal of trouble to avoid confronting the facts about climate change. They tirelessly repeat climate myths, the refutations of which are easily found on websites such as www.skepticalscience.com. These politicians like to say, "I am not a scientist," a truth sadly obvious to any scientist. Yet they have refused to learn what science has discovered about climate change. When Republicans in Congress have held hearings on climate change, they produce tired reruns of political theater. The scientists invited to testify often include the same handful of outlier witnesses whose opinions are known to be compatible with Republican political positions.

Science is the best process that humanity has developed to learn about natural laws. It is self-correcting, based on facts and evidence, not on belief. Marcia McNutt, the distinguished geophysicist who is the president of the US National Academy of Sciences, has said, "Science is a method for deciding whether what we choose to believe has a basis in the laws of nature or not."

Most of the world now accepts that climate science can provide useful input to policymaking. Some 196 countries recently produced the Paris Agreement. Stabilizing the climate and preventing dangerous levels of climate disruption, the goal of this agreement, will require vigorous international efforts and strong American leadership. Only one major country today has an important political party that overwhelmingly rejects climate science. That country is the United States; the party is Republican.

Science shows that the climate system responds to the cumulative emissions of heat-trapping gases. Today's generation thus has its hands on the thermostat controlling future climate. Failure to sharply reduce emissions can lead to sea level rise that will literally change the map of the world. Electing a president who, head in the sand, rejects modern climate science would be risky and potentially disastrous. It would needlessly increase the likelihood that future generations will be condemned to cope with a severely disrupted climate.

5

How to Deal with Climate Change Denying Uncle Pete

This article has been reprinted in the *Bulletin of the Atomic Scientists* several times and must be my best-known article in this publication. It is based on a speech I gave at an annual meeting of the *Bulletin of the Atomic Scientists*.

Perhaps one reason for the popularity of this article is that many people turn out to know an Uncle Pete. It may be a family member or friend, old or young, male or female. The defining feature of this person is that he (or she or they) will not accept any of the research results of mainstream climate scientists. He also loudly informs anyone who will listen that he is invariably right, and the climate scientists are always wrong. Uncle Pete believes climate myths. He may not think that the climate is warming, or if it is, the warming is not large, or the warming has already stopped, or the warming is natural rather than caused by human activities. Uncle Pete may think climate scientists are incompetent or corrupt. He may be convinced that scientists have conspired to silence anyone who dares to disagree with them. In short, Pete will not hesitate to spoil the mood of any event at which he is present. He may seem to be determined to be unpleasant. What can be done about Uncle Pete? I have a few ideas on that subject.

How to Deal with Climate Change Denying Uncle Pete

By Richard C. J. Somerville, December 25, 2018

(This article from the *Bulletin of the Atomic Scientists* is reproduced by permission)

"Birds of a feather flock together," so I am sure that nearly all of those reading this article accept the main findings of climate science. Yet many people do not. Instead, they believe a variety of climate myths.

These include claims that the world is not warming; or if warming is occurring, it is natural and not human-caused; or volcanoes produce more carbon dioxide than we humans do. I know none of you believes these myths, but it seems that almost everybody has an unpleasant relative – call him Uncle Pete – who comes to dinner. Pete spoils the family mood by making these false claims, which he found on talk radio or the internet. I will tell you in a moment why some of the most frequently repeated claims are just plain wrong. I will not have time to cover all of them, and I recommend the website skepticalscience.com for the whole story. It is a collection of the most commonly heard climate myths, and why they are all dead wrong. skepticalscience.com is your key to refuting your own Uncle Pete.

Start with the myth that the warming we have observed in recent decades is natural and not human-caused. First off, let's be clear: The climate has indeed changed naturally in the past, with ice ages being an obvious example. But natural causes simply cannot explain the recent warming. How do we know that? It is very like the story of wildfires, which can be caused naturally, by lightning. But they can also be caused by people, either by carelessness or by arson. And wildfire experts can investigate after a wildfire and determine what caused it. They know how to do the detective work.

We climate scientists are good detectives too. We have discovered what paces the ice ages. It is the slow changes in the Earth's orbit around the sun, which affect how sunlight is distributed over the Earth's surface in the different seasons. Over many thousands of years, these effects are strong enough to cause ice ages to come and go. But over short times, just a few decades, the orbital changes have much too small an effect to produce the observed large warming that has recently occurred.

Through this kind of research, we have quantitatively ruled out all the other natural processes known to affect climate. For example, the Sun powers the entire climate system, and the amount of energy given off by the Sun does vary. But we measure this energy very accurately, and we can demonstrate that its changes are much too small to have caused the observed warming. As for the claim that the extra carbon dioxide added to the atmosphere by human activities is tiny compared to the amounts produced by volcanoes, that too fails quantitatively. Measurements show that human activities, mainly burning coal and oil and natural gas, produce about 100 times more carbon dioxide than volcanoes do.

By burning fossil fuels, we humans have taken over the role of deciding what the climate in coming decades will be. We are no longer passive spectators in the global climate change pageant. We have become the actors. Science provides convincing evidence that the heat-trapping gases produced by human activities are the main cause of the warming observed in recent decades. This aspect of climate science is very firmly established, going back to definitive laboratory experiments in the 1850s. Those experiments showed that carbon dioxide and other gases, present in small quantities in the atmosphere, have powerful heat-trapping properties. In recent decades, the fingerprint of the observed warming, such as how it varies with altitude and geography and season, matches the pattern that we expect from adding heat-trapping gases to the atmosphere. We have found the enemy. He is us.

There are similar convincing refutations of all the other common climate myths. That's why many studies have shown that about 97% of the climate scientists who are most active in publishing research on climate change agree that the observed recent warming is real and serious and overwhelmingly human-caused. Nevertheless, Uncle Pete remains unconvinced. He continues to repeat the myths. You might well ask, "Why is Uncle Pete so stubborn and so resistant to overwhelming scientific evidence?" That's a very good question, and here is my answer.

For many skeptics or contrarians like Pete, the climate change issue is not a science topic at all. For Pete, it is simply an opportunity for the government, and liberals and environmentalists, to make rules and regulations, to interfere with markets, and to diminish the personal freedom of individuals. For Pete, it is just one more way for the authority of the state to control the lives of citizens. This view has nothing to do with science, and no argument based only on science can change it. Uncle Pete, like some actual people I know, may seriously fear that the government will not only decide what kind of car he will be allowed to drive, but will ultimately want to force him to limit his individual carbon footprint, that is, to ration his personal emissions of heat-trapping gases.

Uncle Pete has a high opinion of the free market. He is confident that government actions tend to hinder free markets and thus have the effect of limiting economic progress. He is suspicious of subsidies for renewable energy. He is sure that renewables will never be feasible without big subsidies. Uncle Pete couches his opposition to carbon taxes or fees in statements of this sort: "If you let people keep more of their money, they will invest it in the future." Once again, science is irrelevant here, and no claim that science has shown or proven this or that fact will change Uncle Pete's mind. It is sad but true that most Americans have never met a scientist. Uncle Pete may have his own somewhat strange ideas about how science works and what scientists do. The concept of "peer review" carries no weight at all with Pete; he can easily imagine a corrupt and powerful scientific establishment, conspiring to deny research funding to scientists who disagree with prevailing opinions, and to prevent them from publishing. Pete likes to mention Galileo as an example of an outlier in science who turns out to have been correct. He forgets that Galileos are extremely rare, and that almost everybody who considers himself a Galileo is badly mistaken. Pete may cite eugenics as evidence that the scientific mainstream is indeed sometimes wrong. Pete is very suspicious of us scientists.

Social science tells us that people tend to trust those who share their values and to distrust those who do not. We all know how controversial issues – such as abortion and evolution and gun control – can drive a wedge between people, bitterly dividing this country. And it is high time for us to realize that climate change is a very big "wedge issue" for Uncle Pete. His natural distrust of academics and elites generally is increased if he thinks climate scientists are arrogant people who are scornful of his opinions, who mock his values, and who dismiss his most firmly held convictions.

I urge each of you to engage with the Uncle Pete whom you may know. Have a civil conversation. In his heart, Uncle Pete would probably admit that everybody is entitled to his own opinions, but not to his own facts. When it comes to facts, we scientists have the high ground. The world is warming. It is not a hoax. We measure it. The warming did not stop in 1998 or any other recent year. All the warmest years are recent years. As this is written (early in 2025), 2024 is the warmest year on record globally, 2023 is second, and all of the top 10 warmest years have occurred in the past decade. Natural processes such as El Niño affect year-to-year variations in global average temperature, but the overall trend is clearly upward. Global warming is definitely still continuing.

The atmosphere is warming, and so is the ocean. Sea level is rising. Ice sheets and glaciers are shrinking. Rainfall patterns and severe weather events are changing. Climate change is real, and serious, and happening right here, right now. And it is not natural. Human activities are the dominant cause of the climate changes we have observed in recent decades.

But none of these facts tells us exactly what we should do about climate change. Science can inform wise policy, but it cannot decree or prescribe what the best policies will be. There is no silver bullet, but there is lots of silver buckshot.

In deciding climate policy, science matters, but so do values, priorities, and political convictions. Given the same facts, different reasonable people can easily prefer different policies. For Uncle Pete, attacking climate science and scientists is simply a disguise for what concerns him, which is the prospect of liberals and environmentalists dominating policy, and of a government spinning out of control, a government that in Pete's view seizes power, limits freedoms, increases taxes, regulates markets, and diminishes prosperity.

We do not yet have national agreement on climate change. As you know – and I hope this will not shock anybody – some elected officials in the federal government sound just like Uncle Pete. Despite the strong scientific consensus, climate change policy is contentious politically.

One option for dealing with this political disagreement is to do nothing. Uncle Pete may well favor that option, because it appears to fit well with his sincere conviction that, "if you let people keep more of their money, they will invest it in the future." On doing nothing, I may be able to help Uncle Pete think a bit more clearly.

I am not an expert on energy policy or taxes, but as a climate scientist, I can say something with very high confidence about what will happen if we do nothing. Deciding to do nothing about climate change is like deciding not to have serious elective surgery, such as declining a coronary artery bypass operation that your cardiologist recommends. The operation will involve risks and costs. But declining it will also involve risks and costs, including the risk of a fatal heart attack.

Sadly, we do not have enough conversations about climate change. The media largely avoid the subject, and it was almost invisible in the recent campaign for president. Today, the fact is that we, you and I and the other 8 billion living people, now have our hands on the thermostat that controls the climate of our children and grandchildren. A considerable portion of the carbon dioxide we emit will remain in the atmosphere for centuries and longer. Thus, it accumulates. There is a given allowed amount of carbon dioxide in the atmosphere that we must not exceed if we want to limit warming to any particular target we pick. For the warming target of the Paris Agreement, we are already about halfway to that allowed amount, so we do not have much time left to bring emissions to nearly zero. That's why it is urgent to drastically reduce global carbon dioxide emissions and to do it quickly before we exceed that amount.

It is important to understand that once the world has agreed on a target of how much warming is to be allowed, science can say approximately how much more carbon dioxide can be emitted. The urgency of reducing emissions thus arises directly from the physics and chemistry of the climate system. It has nothing to do with politics.

Mother Nature reacts to the total amount of carbon dioxide. The more carbon dioxide there is in the atmosphere, the greater the climate change will be. If we who are alive today do nothing about climate change, and if the world continues to use the atmosphere as a free dump for carbon dioxide and other waste products of an energy system based on fossil fuels, then we are effectively sentencing future generations to the consequences of a severely disrupted climate. Also, the disruption will not be brief. It will take many thousands of years for the climate to recover after we stop emitting carbon dioxide. So, it is a long sentence. This is not a partisan political statement. It is well-supported, solid science.

The Pentagon, which is certainly not a cabal of liberal environmentalists, takes this issue very seriously, and it has repeatedly characterized unmitigated climate change as a threat multiplier. In the decades and centuries ahead, doing nothing means the world will inevitably see devastating climate change, including agricultural disasters on an immense scale and coastal cities abandoned worldwide, because of sea level increases of many feet.

Vast numbers of people will become environmental refugees, and we will see the destabilization of governments, especially in failed and failing states. In wealthy and powerful countries, such as the United States, governments coping with severe climate change will surely have to act forcefully, including using emergency powers, as in wartime, to preserve order and to minimize chaos and damage. Ironically, in my view, doing nothing about climate change, Uncle Pete's preference, is thus likely to force governments to do exactly what Uncle Pete fears most: seize power and limit freedoms. Doing nothing is a disastrous policy option.

In your civil and mutually respectful conversation with your own Uncle Pete, I hope you can help him think seriously about the prospect of such a horrible, but very preventable, future. We are at a critical crossroads. We still have a chance of limiting climate change to a tolerable level, a level that offers some hope of successful adaptation. Our window of opportunity is still open. But it will not stay open much longer.

We must act. We cannot dither any longer. If Uncle Pete wants to keep the government from controlling his life and diminishing his freedom – as most all of us do – then we all need to learn about and accept the science. We all need to take the threat of climate change seriously. We all must act wisely, and urgently, to minimize that threat and thereby limit the damage of climate change to tolerable levels.

6
Wildfires and Climate Change

This chapter contains a list of 12 points, totaling only a few words, that in my professional opinion are a compact summary of the fundamental findings of climate change science. These 12 points are not an official product of the Intergovernmental Panel on Climate Change (IPCC) or any other entity. I wrote them. However, in my professional judgement as a climate scientist, these points are consistent with the IPCC reports and with many other authoritative summaries of the key findings of scientific research on climate change. I developed this list of 12 points for use in my public lectures. For lectures illustrated with slideshow software, this short list of 12 points has the virtue of all 12 points being brief enough to fit on one slide.

Wildfires and Climate Change

By Richard C. J. Somerville, October 29, 2019

(This article from the *Bulletin of the Atomic Scientists* is reproduced by permission)

Some 2,000 years ago, the great Jewish leader Hillel gave a fine example of brevity in teaching. Hillel was confronted by a skeptic who wanted an explanation of the five books of the Torah. These books are known to Christians as the first five books of the Old Testament. The analyses of these books are long. The skeptic, however, demanded a statement so brief that it could be spoken while standing on one foot. Hillel lifted a foot and said: "Treat others as you would wish them to treat you. That is the entire Torah. The rest is commentary. Now go and study."

The recent terrifying outbreak of wildfires in California, causing extensive damage and massive evacuations, naturally raises the question of connections between climate change and wildfires. Before we speak about such connections, we should make the effort to learn the basics of the science of climate change. I think any rational response to climate change involves first knowing what the facts and evidence are. That is the province of science. We should all learn and accept the facts and evidence, which are objective truth and should be the same for everyone. Nobody is entitled to his own so-called facts.

Fortunately, there is now widespread agreement on the fundamental findings of the science. There will always be outliers in every branch of science, but many studies confirm that about 97% of the scientists who have been most active in carrying out research on climate change and publishing it would agree on the basic research results.

The best summary of climate change science is the assessment reports of the IPCC. Six such reports have been published since 1990 (Note added in 2025: the sixth IPCC assessment report came out in 2021). The physical science portion of the most recent IPCC report is scientifically definitive, but it is long. The physical science part (Working Group One in IPCC language) is more than 2,000 pages long, full of charts and graphs, and not easy reading.

I am certainly not Hillel, science is obviously not religion, and the IPCC reports are not the Torah. But I think the essence of this IPCC report can be summarized in 12 succinct points, which are as follows:

1. It is warming.
2. It is us.
3. It has not stopped.
4. The heat is mainly in the sea.

5. Sea level is rising.
6. Ice is shrinking.
7. CO_2 makes oceans more acidic.
8. CO_2 in the air is 50% higher since the 1800s.
9. It is now the highest in millions of years.
10. *Cumulative* emissions *set* the warming.
11. *Reducing* emissions *limits* the warming.
12. Climate change will last for centuries.

Just 12 points. Only a few words. You can easily speak them while standing on one foot.

One could say much more about each of these 12 points. To start with, note these facts:

1. It is warming. It is not a hoax. We measure it. The atmosphere is warming. So is the ocean. Sea level is rising. Ice sheets and glaciers are shrinking. Rainfall patterns and severe weather events are changing. Climate change is real and serious. It is not a remote threat for the distant future. It is here and now.
2. It is us. We have done the detective work. It is not natural like ice ages. It is human caused.
3. It has not stopped. The warming is continuing. The warmest years on record are all recent years.
4. The heat is mainly in the sea. Over 90% of the heat added to the climate is in the oceans.
5. Sea level is rising globally. The rate of this rise is increasing. The rise is not uniform globally.
6. Ice is melting. Ice sheets on Greenland and Antarctica, as well as glaciers, are all shrinking.
7. Carbon dioxide absorbed by the oceans makes them more acidic. That can affect the marine food chain.
8. The amount of carbon dioxide in the atmosphere is about 50% higher than that in the 1800s, due to human actions.
9. The amount of carbon dioxide in the atmosphere now is the highest it has been in millions of years.
10. Cumulative emissions of carbon dioxide and other heat-trapping substances set the amount of warming.
11. Reducing emissions of carbon dioxide and other heat-trapping substances limits the warming.
12. Climate change, because it takes so long for carbon dioxide amounts to decrease, will last for centuries.

Now we can turn to the connections between climate change and wildfires. A good place to start is a recent IPCC report entitled *Global Warming of 1.5°C*. The unsurprising general finding is that the greater the warming, the worse the consequences of climate change. Thus, a warming of 2 degrees Celsius above preindustrial conditions leads to more severe effects than a warming of 1.5 degrees Celsius. The report cites several scientific studies as "evidence for the attribution of increased forest fire frequency in North America to anthropogenic climate change during 1984–2015, via the mechanism of increasing fuel aridity almost doubling the western USA forest fire area compared to what would have been expected in the absence of climate change."

The rest is commentary. Now go and study.

7
Facts and Opinions about Climate Change

Science is based on facts and evidence. It is not based on beliefs and opinions. Without going into complicated technical details and without any mathematics or scientific jargon, this chapter explains clearly how several important scientific findings about climate change were discovered.

In 2005, at a conference in Aspen, Colorado, I was fortunate to be invited to give a speech on climate change to an audience that included two well-known men who had both been unsuccessful candidates for President of the United States: Al Gore and John Kerry. The next year, I published an article "Medical Metaphors for Climate Issues" (Somerville, 2006), which summarized my speech.

Since then, I have refined and enlarged the concept of medical metaphors, and I have used them in several talks. In this relatively long article, I mention a few examples of medical metaphors. I encourage you to create more of these metaphors.

Metaphors can be powerful. If I say to you that climate science, like medical science, is incomplete but still useful, this comparison brings your entire experience with physicians and other health-care providers, as well as hospitals and vaccinations and medications, into the discussion. You may reflect that, of course, medical science has not yet cured all diseases, or prevented all epidemics, or guaranteed a long life of excellent health for everyone. Yet you know that you and others have benefited greatly from recent advances in medical science. Some diseases have been eradicated. Life expectancy is much longer than it was a few decades ago. Powerful medications such as antibiotics have been developed and made widely available. Devices such as pacemakers have made it possible for many people with severe cardiac problems to live long and healthy lives. Thus, you may be more willing to at least consider the concept that climate science, while imperfect and incomplete, might still be helpful and valid and useful in helping us decide how to cope with climate change.

Facts and Opinions about Climate Change

By Richard C. J. Somerville, December 7, 2020

(This article from the *Bulletin of the Atomic Scientists* is reproduced by permission)

When the *Bulletin of the Atomic Scientists* was founded, climate change science was in its infancy. There were no global climate models, no supercomputers, and no satellite remote-sensing data. Only a few visionaries understood that man-made increases in the amount of carbon dioxide (CO_2) might cause large global climate changes.

It is important to distinguish between facts and opinions. I will first summarize the facts that we have learned from the science of climate change. Then I will give some opinions about what people and governments should do.

I think any rational response to climate change involves first knowing what the facts and evidence are. That is the province of science. There are many indicators measured globally over many decades that show that the Earth's climate is warming. It is definitely warming. It is not a hoax. We observe it and measure it. The atmosphere is warming. So is the ocean. Sea level is rising. Ice sheets and glaciers and snow cover are shrinking. The amount of water vapor in the atmosphere is increasing. It is us because human activities caused all these phenomena. Climate change is real and serious. It is not a remote threat for the distant future. It is here and now.

It has not stopped. The warming is still continuing. We have good estimates of the global average temperature of the Earth's surface from 1880 until the present; the year 1880 was about the time when we first had enough good thermometers located in enough places worldwide to enable us to calculate a meaningful global average. The modern data are the most accurate ones. Over the last 50 years, from the 1970s until now, we know there has been a warming of about 1 degree Celsius, or 1.8 degrees Fahrenheit. All the warmest years on record are the recent years.

The heat is mainly in the sea. Over 90% of the heat added to the climate system is in the oceans. How do we measure the heat stored in the ocean? That's a good question, and it has a fascinating answer.

We now measure this increase in ocean heat content from an array of about 4,000 autonomous floats deployed throughout the world ocean under an international program called Argo. They have no engines and no propellers, but they move with the ocean currents at a depth of 1,000 meters, which is a depth as long as about 10 football fields, laid end to end. That's where they are usually parked, and they are programmed to periodically sink another 1,000 meters lower and then rise to the surface while measuring quantities such as water temperature and salinity. They rise and sink by changing their density. This is accomplished by pumping fluid from an internal reservoir into or out of an external bladder on the float, thus changing the density of the float, because the weight of the float is unchanged while the volume changes.

The floats store the measurements, and then, when they are on the surface, they locate by GPS and transmit the stored data via satellites to scientists. The change in float locations between one transmission and the next provides information on the currents at the depth where the floats were parked. The floats have batteries, which power the communications system and the instruments and the pumps that alter their volume, hence their density, and allow them to rise and sink. As their batteries fail, the floats end their useful lives. They sink to the ocean bottom and must be replaced by new floats. The Argo floats have revolutionized our ability to observe the oceans. Argo data are available to everyone for free in near real time. New floats, allowing sampling to much greater depths, and measuring other variables, are now being developed. Figure 7.1 depicts the global distribution of Argo floats on one day in 2025. Maps, such as Figure 7.1, are updated on a daily basis.

Sea level is rising globally. We measure it from altimeters on satellites. There has been a rise of about 100 millimeters, which is about 4 inches, over the last 30 years or so. The rate of sea level rise is increasing too. Future sea level rise will be much greater than past sea level rise. The sea level rise varies at different locations globally. The local sea level is affected by whether the land at that location is rising or sinking and also by ocean currents, tides, and other factors.

Ice is shrinking. Ice sheets on Greenland and Antarctica and almost every glacier worldwide are shrinking. The GRACE (Gravity Recovery and Climate

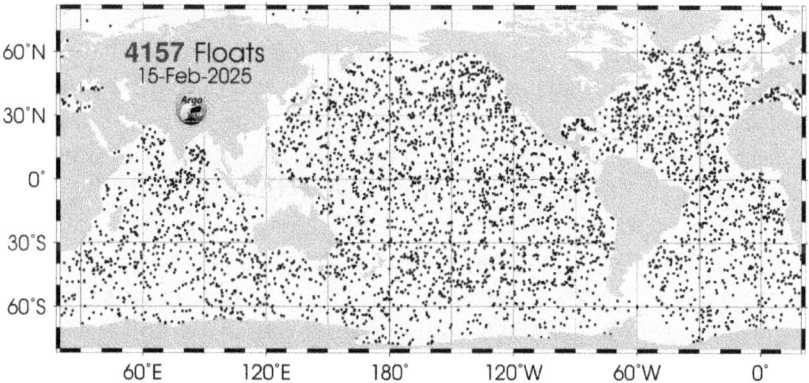

Figure 7.1 Argo floats, showing the locations throughout the global ocean of more than 4,000 autonomous floats on a typical day in early 2025.

Experiment) satellites determine ice mass accurately by measuring the effect of the ice sheets on the Earth's gravity. The technology of the GRACE mission involves a pair of satellites in the same orbit to map the Earth's gravity field by measuring the distance between them precisely. This distance changes when gravity varies, which occurs when passing over ice sheets, and measuring the tiny change in the inter-satellite distance allows scientists to determine the mass of the ice.

Carbon dioxide absorbed by the ocean makes it more nearly acidic, which affects the marine ecosystem and the food chain. The ocean absorbs some of the CO_2 that we emit into the atmosphere. The measurements show that the acidity parameter called pH is decreasing. Seawater is slightly basic (pH is greater than 7), and we observe a shift toward neutral conditions (pH is equal to 7) rather than to truly acidic conditions (pH is less than 7).

The amount of carbon dioxide in the atmosphere, because of human activities, is now about 50% higher than that in the early 1800s. We have good measurements of atmospheric CO_2 over the last 800,000 years. These data come from analyzing fossil air trapped in ice in Greenland and Antarctica. They reveal large variations in CO_2 amounts, associated with ice ages starting and ending. Orbital variations pace the ice ages, causing the CO_2 amounts to change, and initiating a feedback that increases the magnitude of the temperature change.

The amount of carbon dioxide in the atmosphere now is the highest it has been in millions of years. The atmospheric CO_2 amounts in the even more distant geological past, many million years ago, sometimes have been even higher than at present, but the world was a very different place then, which was long before any human beings existed.

Cumulative emissions of carbon dioxide set the amount of warming. The warming caused by CO_2 in recent decades is, to a good approximation, just linearly proportional to the total cumulative amount of carbon emitted into the atmosphere. We do not know exactly how much added CO_2 will produce how much warming – that is, the climate sensitivity question – but we can estimate a range of possible answers to this question, constrained by different kinds of observations. For the middle of the range, 1 trillion metric tons of carbon emitted produces a warming of about 2 degrees Celsius above the temperatures of the early 1800s. We have already emitted about half of this amount. At present, the warming we observe is caused by CO_2 plus several other heat-trapping substances that human activities have also added to the atmosphere. The amount in the atmosphere of these other substances will decrease rapidly when and if their sources are eliminated, but some of the carbon dioxide will remain in the atmosphere for thousands of years. Because of the difference in the amount of time that different heat-trapping substances stay in the atmosphere, carbon dioxide is truly the key "control knob" for climate.

Reducing emissions of carbon dioxide and other heat-trapping substances will limit the warming. We can estimate the cumulative amount of carbon dioxide emitted that would give us a good chance of limiting warming to 2 degrees Celsius (that's 3.6 degrees Fahrenheit) above the preindustrial temperatures of the early 1800s. This is the warming target endorsed by the Paris Agreement of 2015. If emissions had peaked and began to decline several decades ago, then emissions reductions could be gradual, and by 2050 emissions would not yet need to have entirely stopped. Because emissions are still increasing, drastic emissions reductions need to occur quickly and reach zero by 2040.

"Negative emissions," meaning removing some carbon dioxide from the air, are likely to be necessary. This fact illustrates the urgency of acting. Finding a way of removing some of the carbon dioxide is one approach to geoengineering. Here by "geoengineering" we mean the intentional modification of the climate system with the goal of reducing or mitigating climate change. However, nobody has yet demonstrated a way of economically removing large amounts of carbon dioxide from the atmosphere.

The longer we wait before acting, the more drastic the action has to be. The result of failing to act is to increase the likelihood of dangerous climate change.

Because it takes so long for the carbon dioxide in the atmosphere to decrease, climate change will last for centuries. After emissions completely stop, the amount of carbon dioxide in the atmosphere decreases only slowly for several centuries, and about 25% of it remains in the atmosphere for the next 10,000 years or so. The science relevant to this topic is not simple. Several complex processes for carbon removal are involved. The key take-away message is that the climate change caused by adding carbon dioxide to the atmosphere can have very long-lasting effects.

The conclusions I have just recounted are facts. They are fundamental findings from extensive scientific research. They are all well-supported by abundant evidence.

The science is never complete. There is always more to learn. But the science that we have now is already good enough to help us make wise decisions. The many unknowns in the science, such as exactly how fast the Antarctic ice sheet will shrink or exactly how El Niño might be affected, are not the biggest unknowns about future climate. The biggest unknown about future climate is human behavior. Everything depends on what people and their governments do.

The scientific consensus is overwhelming. Climate change is already happening, here and now. About 97% of climate experts – the scientists who are most active in carrying out and publishing research on climate change – agree that the observed recent warming is real and serious and overwhelmingly human-caused, and that it will become even more serious unless we make big changes in how we generate energy. Nevertheless, some people remain unconvinced. They continue to repeat climate myths and falsehoods.

People often ask me, "I am only one person. What can I do about climate change?" Here is my answer. We need to persuade more people that this problem is serious. Governments tend to respond when enough people become concerned, and when they vote their concerns. I urge everyone to engage with people you may know – family, friends, colleagues – who do not accept the fundamental findings of climate science. Explain to them the facts you have learned about our changing climate. Listen to them respectfully and carefully. Be alert to the common climate myths and falsehoods that they may think are true. If you see something, say something. Have a civil conversation. Have many conversations. In their hearts, almost all of us would surely agree that everybody is entitled to his own opinions, but not to his own facts. And it is science that supplies the facts about climate change.

We humans have become the dominant actors in causing the rapid climate change we now observe. Human actions now overwhelm all the natural processes. This may seem counterintuitive, but it is true. You and I, and all the people who are alive today, now have our hands on the thermostat (Figure 7.2) that controls the climate of our children and grandchildren. Metaphors such as the thermostat can be superb communication tools.

Think about medical metaphors. Here are a few: We climate scientists are planetary physicians. Climate science and medical science will both always be imperfect and incomplete, but both are already very useful.

Figure 7.2 The thermostat is a powerful metaphor. Human activities are able to alter the climate, just as a thermostat is able to alter the temperature of a building. People who are alive now effectively have their hands on the thermostat that controls the climate of our children and grandchildren. Photo credit: Richard C. J. Somerville.

7 Facts and Opinions about Climate Change

When your doctor tells you to stop smoking and lose weight and exercise more, you do not argue with her. You do not call her a radical alarmist. You do not ask her to name the date when you will have a heart attack.

Physicians have advanced academic credentials and many years of training and experience. We climate scientists also have the same. We are not conspiring to fool people. Do you really think your doctor is a crook? She is not. Neither are we.

A fever of only a few degrees can indicate a serious disease. Global warming is just a symptom of planetary ill health, like a fever.

Prevention is better than cure. Quitting smoking, like quitting using fossil fuels, is not easy to do. And the main benefits of quitting come in the long-term future.

Choosing to have major surgery involves cost and risk. People know that choosing to do nothing also has costs and risks.

The laws of climate science and medical science are all immune from political tampering. You cannot fool Mother Nature. Mother Nature always bats last.

Here's an effective metaphor. Imagine you are watching a major-league baseball game. The slugger who is thought to be on performance-enhancing drugs hits a home run. The person next to you asks, did the steroids cause it? That's really the wrong question. You cannot be sure they caused it, because he was already a big-league slugger when he was clean. And even with the drugs, he can still strike out now and then. But at the end of the season, you see in his statistics that he hit more homers than he used to. The steroids (Figure 7.3) increase the odds of home runs. Climate is the statistics of weather, and carbon dioxide is the steroids of climate. It changes the odds. The odds are higher now for all sorts of extreme weather, because climate change has altered the environment in which all weather occurs.

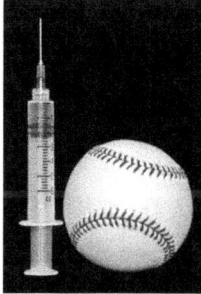

Figure 7.3 Carbon dioxide is the "steroids" of climate, increasing the odds of extreme weather events, just as "steroids" can increase the odds of exceptional athletic performance.

This metaphor works for other sports too. For example, baseball is not popular as a spectator sport in France, although it is played in many schools, but professional bicycle racing is very popular in France. As a result, many French people know that a bike racer on drugs will not win every race, but the drugs do change the odds and increase his probability of winning.

The main barriers to action on climate change are a lack of widespread political will and a lack of wise and inspiring leadership. Science can help to inform policy, but only concerned people and responsive, capable governments can first decide what policies are best and then implement them. Today, despite a strong scientific consensus, climate change is controversial politically.

We do not have to accept a future with devastating climate damage and disruption. If we continue to use more and more fossil fuels to generate the world's energy, we will be sentencing our children and grandchildren to many centuries with a severely damaged climate and great suffering. In your conversations, try to help people understand that this bleak future is entirely preventable.

Faced with these threats, almost all the nations of the world agreed in Paris in late 2015 to limit the warming to a specific maximum amount. That amount is 2 degrees Celsius, or 3.6 degrees Fahrenheit, above the average global temperature in the early 1800s, before human activities began to have a large effect.

After the Paris Agreement, is the glass half-empty or half-full? I am guardedly optimistic, for these reasons:

- World leaders are now engaged; at least almost all of them are.
- Emissions of heat-trapping gases have begun to decline in some places.
- Solar and wind energy are getting cheaper every year.
- Renewable energy use is increasing rapidly.
- Many corporations are now acting to reduce emissions.
- State and local governments in the United States are acting too, despite federal inaction.
- Many other countries are showing rapid progress.

Recent polling shows that in the United States, many more people accept the science and are very concerned about global warming or climate change than was the case only a few years ago. However, the issue has become extremely partisan. Recent polls show that the substantial increase in the number of Americans calling climate change a top priority has been limited to the Democratic party side of the political spectrum. However, about 80% of Republicans, including virtually the entire leadership of the Republican party, have not changed their minds and still reject the science. We have a long way to go. I think we should keep climate change science separated from climate change policy.

There is no silver bullet that solves all the challenges of climate change, but there is a lot of silver buckshot, including increased energy efficiency and energy conservation, and much more use of sun, wind, and water to provide the energy the world needs. These renewable resources are widely available now and already cost-competitive with fossil fuels. We have the technology, and it is improving. In the United States, even without energetic action by the federal government, I am guardedly optimistic.

Market forces now favor carbon-free energy. Coal companies are going bankrupt. Solar and wind energy without subsidies are in many cases already cheaper than fossil fuels. Electric vehicles are happening fast. Much energy policy in the US is set at state and local levels, not in Washington.

Research suggests that messages that may invoke fear or dismay are better received if they also include hope. We should include positive messages about our ability to solve the problem. We can explain that future climate is in our hands.

Politics and priorities and values do have a role to play in deciding which actions are best, but any rational policy begins by accepting the science. Once again, people are entitled to their own opinions, but not to their own facts.

This article has been written in 2020 during the global coronavirus pandemic, and I have a few gentle words to say about some climate change lessons that we might learn from the pandemic now gripping the entire world.

One obvious point is that climate change science, like coronavirus epidemiology, is incomplete, still developing but already extremely useful. In both domains, we have learned we can trust scientists more than politicians or pundits or anybody else who is not really an expert on the science of the subject, whether the subject is climate change or infectious diseases. We have also learned that the challenges in both climate change and the pandemic are global. The entire world is affected. The solutions have to be global too.

The pandemic also illustrates the wisdom of the statement that, "Everybody is entitled to his own opinions, but not to his own facts." The facts about climate change, and about COVID-19, are objective truth, and they should be the same for everybody, regardless of people's ideology or politics. When it comes to making policy, sound science can inform wise policy. However, policy can also depend on many other factors, such as people's priorities, their convictions about economics, what they regard as the proper role of governments, their risk tolerance, and, of course, public opinion. That's true for meeting the challenge of climate change, just as for meeting the challenge of COVID-19.

The pandemic reminds us how valuable science and scientists are. The recent discussions in the news – such as about how clinical trials of drugs and vaccines work – are very educational. The medical scientists who develop new medicines do their best to make sure they are safe and effective, and they will not release them for widespread use to the public until they are absolutely convinced of that. They are real experts and are very careful. So are climate scientists.

Always remember why we want to have conversations about climate change science. We want to inform people. We want to motivate them. We want them to act.

8
Weaning a House and the World from Fossil Fuels

Lessons Learned

Frequently, when I have given talks about climate change, audience members have asked me what they can do themselves, and what I have done myself, to help limit emissions. I have discussed with several scientific colleagues about what they have done themselves to reduce their own emissions of carbon dioxide. The different approaches they have taken ranged widely. Some of my colleagues have drastically reduced their airplane travel. Others have bought various fuel-efficient cars, including electric cars and hybrids. My family and I decided to replace a furnace in our house, one that burned natural gas, with an electric heat pump system. We also installed a solar photovoltaic system including an array of solar modules or panels on our roof. In that way, we could generate most of the electricity used by our house from renewable solar energy. Also, after the reduction in electricity costs became large enough to pay for the solar photovoltaic system, the electricity would be free. This article summarizes what we did and what we have learned.

Weaning a House and the World from Fossil Fuels: Lessons Learned

By Richard C. J. Somerville, August 4, 2022

(This article from the *Bulletin of the Atomic Scientists* is reproduced by permission)

As a climate scientist, I recently carried out two interesting research projects. Both are relevant in the efforts to reduce the danger of human-caused climate change. First, I decided to investigate how a family in the United States might quickly modify its house to reduce its emissions of carbon dioxide and its consumption of fossil fuels to near zero while at the same time saving money. I picked my own family and our house as the guinea pigs for this experiment. Second, I worked to increase my understanding of whether the world as a whole could rapidly reduce global emissions of heat-trapping substances. I discovered an assessment that I found persuasive in a book by Vaclav Smil, *How the World Really Works: A Scientist's Guide to Our Past, Present and Future* (Smil, 2022).

It took only a few months to modify my house to operate almost entirely on electricity and to generate nearly all that electricity from solar energy rather than fossil fuels. The solar photovoltaic system we installed is on track to pay for itself in only a few years. Weaning the entire world from fossil fuels, however, is a staggeringly complex and difficult task. Imagining a better world for tomorrow is relatively easy. Getting to that better world, starting from the existing world of today, is not easy at all. Vaclav Smil makes the case that this task will require immense changes in many areas where fossil fuels are now vital to the production of massive amounts of materials indispensable to modern civilization. He estimates that completing that transition will likely take several decades or even longer.

My family occupies a single-family home in southern California. The house was built in 1978. We have owned it since 1979. It is a one-story, three-bedroom house with a total area of about 2,700 square feet or 250 square meters. It is a frame house on a concrete slab with a white stucco exterior and a tile roof. It has an attic, used mainly for ducting, but no basement. The climate of this area is formally classified as Mediterranean, with hot, sunny, dry summers and mild winters. Winter is cooler and wetter than summer, but total precipitation is lower than in many typical Mediterranean climates. Annual precipitation in our area is usually less than about 12 inches or 30 centimeters.

Like many houses in North America, our house had been heated in winter by a forced air system, using a furnace fueled by natural gas. The house was not originally air conditioned, but it is only a few kilometers from the Pacific Ocean. Breezes from the ocean and a lagoon provide a significant amount of natural cooling. However, in recent years, summers in the region have become somewhat warmer and we decided to air condition our house.

To reduce our energy consumption, we chose to remove the furnace and install a modern reversible electric heat pump system. This system heats the house in winter and cools it in summer, with three zones, independently controlled. The system components included new ducting, air handlers and purifiers, and electronic thermostats. Heat pump systems have many advantages over traditional home heating systems based on combustion of fossil fuels. They are more efficient, can both heat and cool, improve air quality, are long-lasting, require little maintenance, are quiet, and provide health benefits as well as comfort. On the other hand, they are complex, often have high up-front costs, and are not ideal in all areas. Our heat pump system cost about $39,000. It was installed in August 2021.

Our house is located near San Diego, California, at a latitude of about 33 degrees North. There is abundant sunshine throughout the year, and the house has a large unobstructed south-facing roof. Thus, it was a good candidate for a solar photovoltaic system to generate electricity. In 2021, a solar energy contractor designed a system with 26 solar modules or panels on our roof. The system included 26 modern microinverters to convert the direct current produced by the solar modules to the alternating current required by the house. The system also included a necessary upgrade to the main electric service panel of the house, with safety features required by local codes.

The estimated annual energy production of this solar photovoltaic system is about 16 megawatt hours (MWh). This amount is also approximately the estimated annual consumption of electricity by the house. The solar system produces significantly more electric energy in summer than in winter, because the Sun is higher in the sky in summer, and summer days are longer. The contractor's software had predicted that the system would produce about 1.0 MWh per month in winter and about 1.6 MWh per month in summer. The system included detailed monitoring capability, and the first few months of data showed that the actual production of electric energy was very close to these predicted amounts.

During times when the solar photovoltaic system produces more energy than the house is using, it sends the excess energy into the grid, and the electric utility credits us for this electric energy. Under current California regulations, this solar photovoltaic system is estimated to pay for itself in about six years, after which the electricity generated by the solar system will be essentially free. However, such estimates are often unreliable, because experience suggests that both the California regulatory regime and retail electricity prices can change substantially over short periods.

One highly unpredictable quantity is the future price to be credited by the utility to the homeowner for excess energy fed into the grid. Considerable uncertainty also characterizes the future price to be paid by the homeowner to buy electricity from the utility, during times when the solar photovoltaic system produces either no or less electricity than the house is using. Of course, the utility always has to have some excess generating capacity ready to handle any increase in demand. However, on annual average, our solar photovoltaic system generates about as much electricity from the Sun as our house uses.

This solar photovoltaic system was turned on and began producing electricity in January 2022. Its total cost was about $43,000. We recognize that many families could not afford the cost of the heat pump system plus the solar photovoltaic system. However, US law in 2022 provides a federal income tax credit of 26% of the solar system cost. This credit will reduce our 2022 income tax by about $11,000, so our net cost for the solar photovoltaic system will be about $32,000. At the time I wrote this piece, the federal income tax credit was set to decrease from 26% to 22% in 2023 and to expire in 2024, but subsequent legislation has changed this.

My next goal was to better understand how modifying our house relates to reducing the risk and magnitude of human-caused climate change. A relevant resource is the work of Vaclav Smil, whom I first met in 1995 in Aspen, Colorado. He was a professor at the University of Manitoba in Canada. He had come to Aspen at the invitation of the Aspen Global Change Institute (AGCI) to participate in a workshop relevant to mitigating climate change. I was impressed with Smil's breadth of knowledge and his ability to speak rapidly without notes. Listening to Smil or reading his books is, metaphorically speaking, "drinking from a fire hydrant." One is deluged by data. Smil spoke at AGCI workshops in Aspen several times.

Smil was born in 1943 in a mountainous and forested area now in the Czech Republic. At the time of his birth, however, this area was part of the German Protectorate of Bohemia and Moravia. After World War II, the prewar state of Czechoslovakia was reestablished. It quickly became a Marxist–Leninist country dominated by the Soviet Union. In August 1968, a period of liberalization in Czechoslovakia ended violently when the country was invaded militarily by the Soviet Union, assisted by several other Warsaw Pact countries. (The military invasion of Ukraine by Russia in 2022 has several similarities to this 1968 event.)

Smil had studied for five years at the Faculty of Natural Sciences of Charles University in Prague, attending a large number of courses in a wide range of subjects and earning an undergraduate degree and a graduate degree, the latter approximately equivalent to the American degree of Master of Science. Smil, who had refused to join the Communist party, and his wife, Eva, a recently graduated physician, left Czechoslovakia in 1969, shortly before the Soviets closed the country's borders, emigrating first to the United States. After arriving in the United States, Smil quickly earned a doctorate in geography in 1971 from the Pennsylvania State University.

In 1972, Smil accepted the first job offer he received and joined the faculty of the University of Manitoba, located in Winnipeg, Canada, where he has remained for some 50 years. Reputed to be somewhat reclusive, Smil does not own a cell phone, does not like to give interviews, and does not go to faculty meetings. He reads prodigiously; I have heard him say that he used to read about 100 books per year, but the number has recently dropped to about 75. His memory is extraordinary.

Smil's productivity over his half-century career at Manitoba has been astonishing. He has published some 500 papers and more than 40 books. His books are mainly intended for the general reader, and they range widely, but energy has been a dominant theme in many of them. In 2022, he published *How the World Really Works: A Scientist's Guide to Our Past, Present and Future*, a magisterial and up-to-date overview of Smil's work on energy, with an emphasis on the climate change issue. Of the more than 300 pages in this book, about 70 are devoted to detailed references and notes, including the sources of Smil's data.

Vaclav Smil knows that human-caused climate change is an important and serious problem that should come as no surprise to anyone. Citing work in the 1800s by Eunice Foote, John Tyndall, and Svante Arrhenius, Smil scolds the world for ignoring what these pioneering scientists had discovered: "Clearly, we did not have to wait for new computer models or for the establishment of an international bureaucracy to be aware of this change and to think about our responses."

I think Smil may be a little too harsh in this reproach. One should recognize the long time it took for the work of these early pioneers to be accepted by the scientific community. In 1958, when I entered Penn State as a freshman meteorology major, climate change science was truly in its infancy. At that time, only a few scientists clearly understood that man-made increases in the amount of atmospheric carbon dioxide might cause large climate changes. In 1958, there were no global climate models, no supercomputers, no long-term record of monitoring atmospheric carbon dioxide amounts, and no satellite remote sensing data.

The definitive one-volume summary of atmospheric science when I began my university education was the *Compendium of Meteorology*, a multiauthored book of more than 1,000 pages, published in 1951 by the American Meteorological Society. The *Compendium* article on climate change, written by the distinguished British climatologist C. E. P. Brooks, reflects the prevailing expert opinion of that time. That article began by stating that variations in atmospheric carbon dioxide amounts had been considered by some scientists as a cause of climate change, "but the theory was never widely accepted and was abandoned when it was found that all the long-wave radiation absorbed by CO_2 is also absorbed by water vapour." In the recent years, spectroscopy (the study of the absorption and emission of radiation by matter) has greatly advanced, and today we know this assertion by Brooks is simply wrong. Brooks went on to state that a recent increase in the atmospheric carbon dioxide amount due to burning coal had been cited "as an explanation of the recent rise in world temperature. But during the past 7,000 years there have been greater fluctuations of temperature without the intervention of man, and there seems to be no reason to regard the recent rise as more than a coincidence. This theory is not considered further." There is a valuable lesson about humility here.

In the 60 years or so since 1961, when I graduated from Penn State with a bachelor's degree in meteorology, the field of climatology, and our understanding of climate change have truly been revolutionized. Sixty years ago, as the *Compendium* article (Brooks, 1951) demonstrates, very few scientists thought that changes in atmospheric carbon dioxide amounts could affect climate. Today, we understand that Brooks was wrong. Climate is indeed sensitive to carbon dioxide. To limit climate change, there is an urgent need to greatly reduce emissions of carbon dioxide and other heat-trapping substances. If the world wants to avoid making climate damage more severe, the amount of carbon dioxide in the atmosphere must stop increasing. That means global emissions of heat-trapping substances like carbon dioxide must go to zero and stay there, because some of the carbon dioxide emitted into the atmosphere remains there for millennia.

Smil shares this sense of urgency. However, he has a low opinion of climate models and of long-range predictions and models in general. I think Smil may not fully understand that models are useful for more than making forecasts. They also play a central role in producing nearly all the climate observations that scientists use. For example, models derive atmospheric temperatures from the radiances that satellites measure. They also aid in exploring aspects of climate change other than warming.

In several important ways, I think Vaclav Smil resembles the economist, Thomas Piketty. They both use a wide variety of data to help us understand the world. Piketty draws on nineteenth-century novels, for example, because the writers Honoré de Balzac in France and Jane Austen in England understood that if you knew how much land a farmer owned, you could estimate his income accurately. Piketty's book *Capital in the 21st Century* is his masterpiece.

Both Smil and Piketty are big-picture analysts. They want to know how the important aspects of their respective worlds really function, and they are realists. Others can analyze Smil's thinking because he provides the sources of the data he works from. I suspect Smil is quite right overall, and many climate activists who talk about weaning the world quickly from fossil fuels rarely understand the details of how crops become food and how steel and cement are used in cities and highways, or how widespread plastics have become and how central fossil fuels are to these processes. Smil's value as a thinker about climate change is that he strives to be a reality check. He wants us to learn how to recognize nonsense and filter it out.

Mundane but very important aspects of human endeavor, such as agriculture, producing steel, and making concrete, are central to Smil's argument. Smil calls ammonia, steel, concrete, and plastics, "the four pillars of modern civilization." Countries not yet fully developed now crave these pillars, too. All four pillars require very large amounts of fossil fuels to produce. Thus, Smil seems pessimistic about prospects for the world weaning itself rapidly from fossil fuels. However, he would surely describe his view as realistic rather than pessimistic. Smil writes: "Modern economies will always be tied to massive material flows, whether those of ammonia-based fertilizers to feed the still-growing global population; plastics, steel, and cement needed for new tools, machines, structures, and infrastructures; or new inputs required to produce solar cells, wind turbines, electric cars, and storage batteries. And until all energies used to extract and process these materials come from renewable conversions, modern civilization will remain fundamentally dependent on the fossil fuels used in the production of these indispensable materials."

Although I do not agree with everything Smil has to say, I do enthusiastically admire his work and highly recommend this book to everyone interested in climate change. Smil repeatedly emphasizes his conviction that weaning the world from fossil fuels will take much longer and will be much more difficult than most people realize. He advocates many common-sense steps, such as reducing food waste and driving smaller cars. The title of his book, *How the World Really Works*, conveys his conviction that we are being extremely naive if we continue to insist, as many of us are prone to do, that we already have all the necessary technology, so we are now fully able to limit climate change to moderate or tolerable amounts, but we lack only the political will to do it. Smil tells us that ending the world's addiction to fossil fuels will be difficult and will take a long time, even if we develop the political will to act.

Modifying my family's house so that it runs entirely on electrical energy, and then providing that energy "for free" by putting solar modules on the roof, has been educational for me. It has helped me to appreciate Smil's central point. He makes the case for how extremely difficult it will be, and how very long it is likely to take, to wean the entire world, not just one house, from fossil fuels.

Notes to Chapter 8

[1] Bill Gates published his book *How to Avoid a Climate Disaster* in 2021. In it, he writes, "I had become convinced of three things:

1. To avoid a climate disaster, we have to get to zero greenhouse gas emissions.
2. We need to deploy the tools we already have, like solar and wind, faster and smarter.
3. And we need to create and roll out breakthrough technologies that can take us the rest of the way."

Gates's prescriptions thus include doing everything we already know how to do, but doing it better, and they additionally include developing many "breakthrough technologies" that do not yet exist. Indeed, Gates is spending some of his own vast wealth in efforts to speed up the process. Gates also states, "I think more like an engineer than a political scientist." Gates is not an engineer, but he does have great faith in technology, science, and reason. He also is known to admire Vaclav Smil. However, Gates gives no details of who will pay the bills for the efforts he advocates, or how to persuade the many reluctant governments and corporations to act.

Mark Z. Jacobson of Stanford University and his coauthors have a long history of describing and advocating solutions to the climate change challenge based on using only the three main renewable energy resources: wind, water, and solar. There is a section on "political uncertainty" in their 2022 paper, "Low-cost solutions to global warming, air pollution, and energy insecurity for 145 countries." This paper is cited as DOI: 10.1039/d2ee00722c and is also available at http://web.stanford.edu/group/efmh/jacobson/Articles/I/145Country/22-145Countries.pdf.

The section on political uncertainty does state that "the scale of the transition is enormous" and the pace required depends "on whether manufacturing and deployment can be ramped up fast enough." However, no details are given as to how to bring about the required changes. I think Vaclav Smil might well respond by noting that, like many other authors, Jacobson and his coauthors describe a desirable destination but cannot provide a detailed and realistic map for reaching it.

Vaclav Smil himself has published an extensive criticism of an earlier book by several authors, including the physicist and prolific writer Amory Lovins, who has long been associated with the Rocky Mountain Institute. This criticism appeared in Smil's paper, "Rocky Mountain Visions: A Review Essay." This paper was first published 27 January 2004 and is available at https://onlinelibrary.wiley.com/doi/abs/10.1111/j.1728-4457.2000.00163.x.

Smil's paper is a review of the 1999 book by Paul Hawken, Amory Lovins, and L. Hunter Lovins, *Natural Capitalism: The Next Industrial Revolution*.

In this paper, Smil attempts not only to critique this book but also to address the general category of optimistic prescriptions for a better future, prescriptions that he considers naive or unrealistic, because they do not acknowledge or consider the obstacles and difficulties that Smil would later summarize in his 2022 book, *How the World Really Works*. In this 2004 review essay, he says, "I share their calls for technical rationality, higher efficiency, lower environmental impacts, and more considerate farming – but I cannot foresee such easy walks and such stunning rewards in so short a time as they claim or imply. My quarrel is not with their goals but with the excessive promises, repeated exaggerations, wishful thinking, and righteous insistence that theirs is the only enlightened way."

PART II

Understanding and Communicating Climate Change Science

Introduction to Part II

I am convinced that an important aspect of the climate change challenge is simply that too many people rarely or never talk about it. Any issue that we refuse to talk about can quickly become one that we will do nothing about. We also know that discussing a problem can often help in solving it. I suggest that when reading Chapters 9–12, you might consider the benefits of communicating more about climate change with people you know.

The material in Chapters 9–12 is based mainly on *Communicating Climate Change Science*, written by Richard C. J. Somerville and published in the first edition of the open-access digital textbook *Bending the Curve: Climate Change Solutions* (Ramanathan, 2019), Copyright © 2019 by the Regents of the University of California. This material is reprinted here by permission under a Creative Commons Attribution 4.0 International license (CC BY 4.0). This license allows anyone to share, copy, redistribute, and transmit the work for any purpose, even commercially, as long as they credit the original author and publisher for the creation of the work. This Creative Commons license is recommended for maximum dissemination and use of licensed materials.

9
Preparation

Preparing well is the first step to understanding well and then communicating well. Preparation includes knowing general principles of communication and having access to valuable resources. It also includes acquiring an adequate knowledge of the science of climate change.

Susan Joy Hassol is a communication expert. She and I spent many years working together in communicating climate change science (Somerville and Hassol, 2011). She maintains the website www.climatecommunication.org. It contains much valuable material, including detailed information about the key lessons we have learned. This is a website I strongly recommend to you. It is a rich resource for information on communicating climate change science. Here in 12 words is the guiding philosophy that underlies our approach to climate science communication: *Use simple clear messages, repeated often, by a variety of trusted messengers.*

Many people, when attempting to communicate complex subjects, typically fail to craft simple, clear messages and repeat them often. Instead, they overdo the level of detail, so people have difficulty sorting out what is most important. In short, the more you say, the less they hear. Climate scientists often fall into this trap when trying to explain what they have learned to the broad public. They know a lot, so they want to say a lot. That is a mistake. Think about the experts in various fields whom you may know, such as your doctor. The doctor has spent many years learning a great deal about medical science, but only a very foolish doctor would try to tell you everything relevant to your health that medical science has discovered. Instead, a wise doctor speaks to you in simple clear terms.

I think that those who have studied this subject most seriously and carefully have now awakened to the complex challenges of communicating about climate change, when much more than the science is at issue. Our awareness now includes cultural and psychological issues.

Still, most people say they need more information about the science, so scientists and others are challenged to deliver scientific information in more accessible and effective ways. Much of what I summarize in this short chapter is based on the resources available on the website www.climatecommunication.org. This summary reflects the ideas and recommendations on that website for combining accurate science knowledge with effective techniques for communicating with your friends and colleagues and also with the public.

Trusted messengers can have an enormous impact and can motivate people to bring about change. Think of Mahatma Gandhi, or Nelson Mandela, or Martin Luther King, Jr. A privilege for me and a memorable day in my life occurred in 2012 when my Scripps colleague Professor Veerabhadran "Ram" Ramanathan and I spent a day on a stage at our university, UC San Diego, conversing with the Dalai Lama about climate change, with thousands of students in attendance, and many more people watching on television. You do not have to be a Tibetan Buddhist to understand that when the Dalai Lama speaks, millions of people listen. The Dalai Lama, like the Pope, is respected and revered worldwide, regardless of people's religious convictions (Figure 9.1). The Dalai Lama is an excellent example of a trusted messenger whose statements about climate change can profoundly affect public opinion.

Climate change is much more than a scientific topic. I am convinced that confronting climate change is fundamentally a moral and ethical issue. It involves considerations of intergenerational equity. What do we owe to people

Figure 9.1 Richard C. J. Somerville and the Dalai Lama in a public conversation about climate change on April 18, 2012, at the University of California, San Diego. Photo credit: Sylvia Bal Somerville.

who will come after us? Speaking as just one citizen of the Earth, I suggest that, at a minimum, we owe our descendants a planet that is as undamaged as the one we inherited from previous generations. It is also a matter of equity. What do we in the rich nations owe to the billions of people now alive who do not yet enjoy what we would consider a bare minimum of rights and privileges? These include adequate food, access to clean water, decent health care, education, security, and, not least, the material comforts that come from a certain level of affordable energy. Our own prosperity has been built on having such energy, but we have used the atmosphere as a free dump for the waste products from our energy system, such as carbon dioxide (CO_2). We now realize these waste products can produce horrific side effects. And finally, what do we human beings owe to the natural world, now threatened with unprecedented levels of species extinction?

Scientists and everybody else can improve their communication skills by considering their audience, knowing who the audience members are, and learning what they care most about. Why is climate change important to them? This approach to communication often means emphasizing impacts of climate change happening now, here in our own backyards, rather than impacts far away and in a distant future. It can also mean making connections between climate change and what many people are experiencing in their daily lives, such as increases in extreme weather.

In addition to knowing your audience, it is important to know yourself. Analyze your own strengths and weaknesses as a climate change communicator, both in general terms and for each audience you face. Showing that you are interested in what your audience cares about, and showing that you are a warm, likeable, knowledgeable, and trustworthy person, can make you a much more effective communicator. Seek feedback from your audience. Learn what others think of your abilities as a communicator of climate change science.

Knowing your subject matter is crucial. Being well informed about the science of climate change is an obvious step in preparing to communicate it. For example, you can and should learn the most common myths and falsehoods about the science, and you can be prepared to refute them convincingly. Become something of an expert yourself first, at least in certain areas of climate change science, and only then try to communicate what you have learned. When answering a question, if you do not know the answer, say so. Do not guess. You are not expected to know everything.

Facts matter. Here are some facts: The world is warming. It is not a hoax. We measure it. The warming has not stopped. All the warmest years on record are recent years. The evidence for warming is not a weak thread. It is a strong rope. The atmosphere is warming. So is the ocean. Sea level is rising. Ice

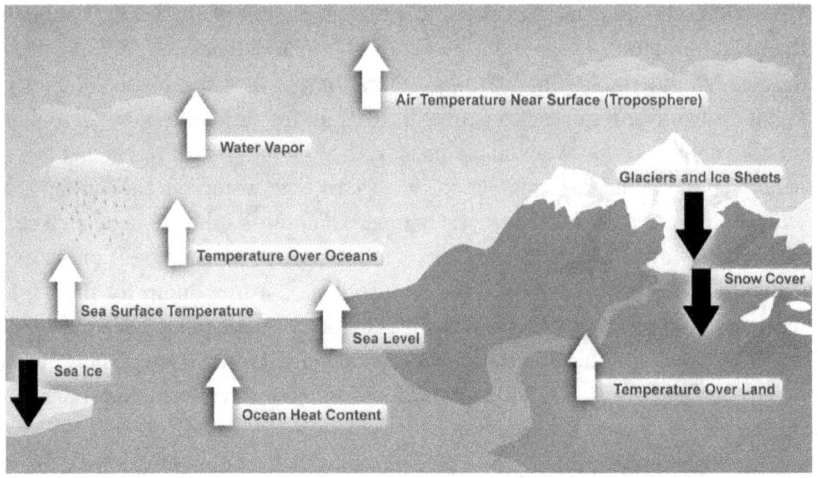

Figure 9.2 Observational evidence for global climate change. White up arrows indicate increasing trends in a warming climate. Black down arrows indicate decreasing trends in a warming climate. All of the observed trends are in the same direction as the arrows. They all support the conclusion that the climate is warming.

sheets and glaciers are shrinking. Rainfall patterns and severe weather events are changing. Climate change is real and serious. It is not a remote threat for the distant future. It is here and now. Figure 9.2 depicts the observational evidence that our climate is indeed warming.

All of the observed changes are consistent with a warming climate. Climate science communicators should be familiar with these aspects of the findings of climate research, and they should know important details about how these observations are made and why they are trustworthy. The best summary of climate change science is the assessment reports of the Intergovernmental Panel on Climate Change, or IPCC. In Chapter 6, I introduce a list of 12 brief points that I have developed to summarize some of the most important findings of climate change science. These are scientific results, well supported by extensive research and endorsed by every relevant major scientific organization in the world.

10
Stories

Stories are a wonderful way to engage the audience. Scientists are widely admired, and scientific research has clearly brought many benefits to humanity. You might naively think that most people would be inclined to accept the main findings of climate science. Yet many people strongly disagree with climate scientists. Some people insist on frequently repeating several climate myths and falsehoods.

To climate scientists, whose goal is to discover the truth about climate change, it really does not matter whether or not some people find the science believable. Science is based on facts and evidence, not on beliefs. Most people have a very vague conception of what science is and what scientists do. People often recall their high-school chemistry course, for example, as a boring exercise in memorizing useless material, such as the periodic table of chemical elements, and then forgetting it as soon as possible after the exam. Most people have never met a scientist. It is not just that some people do not know elementary facts, such as that the Earth goes around the Sun once a year. It is that they may have no idea how such facts were discovered. There is nothing wrong with belief. For example, it is important for people to believe that it is good to treat other people well. But in some domains, there is another way to find out what is true. That is to compare one's beliefs with facts and evidence. Science is the name we give to doing that.

We climate scientists sometimes think of ourselves as planetary physicians. I had a fascinating experience not long ago. My primary care doctor retired. I had to choose a new doctor. When we met for the first time, my new doctor said, "Sit down. Let me tell you how I practice medicine. First, I am competent. I know what I am doing. Second, I am honest. If there is something I do not understand, I will tell you. Third, I am here only to advise you. You will make all the decisions."

I was impressed. No doctor had ever talked to me like that. We climate scientists are planetary physicians. We are also competent and honest and here only to advise. We have learned many things about climate, but we still have a lot to learn. Like the findings of medical science, our understanding of climate, incomplete though it is, is already highly useful.

For example, the fundamental question – whether all of us, some 8.2 billion humans as of early 2025, have caused the world to warm up in recent decades – has already been answered. The answer is yes. We have settled that issue. At least, an overwhelming majority of the most active climate scientists involved in research consider it settled. Some other people may choose not to believe it. There are people, like the fictitious Uncle Pete in Chapter 5 of this book, who are unwilling to believe things that they wish were not true, or who just do not trust experts.

The public has come to respect medical science, however. Although there will always be gullible people, most of us know that there is a difference between real experts and charlatans. Most people will not listen to, or act on, medical advice from a quack who can talk about medicine but who is not really a physician. Everybody accepts this situation. Even the least enlightened members of Congress do not hold hearings to denounce modern medical science as a hoax. Yet, a few politicians and others do denounce climate science in exactly this way.

Medicine is different. At your annual checkup, if you are sensible, when the doctor tells you to lose weight and exercise more, you do not argue. You do not complain that medical science is imperfect and cannot yet prevent cancer or cure AIDS. You do not label your doctor a radical alarmist. You know, and your doctor knows, that medical science, while imperfect and incomplete, is still good enough and valuable enough to provide advice well worth following.

Of course, some people just do not do what experts tell them. Not everybody takes the medications their doctor prescribes. "Noncompliance" by some patients can be a serious problem for physicians. We should keep all this in perspective. Lest we fall into the trap of thinking that medical science is a perfect role model for us climate scientists who crave more public esteem, it is also good to remember that it took a long time for many medical results to acquire widespread acceptance. Some scientists in the 1930s already suspected that tobacco caused cancer. The evidence was widely known to be strong by the 1960s. Yet the high-profile anti-tobacco lawsuits in the United States began only in the 1980s, about half a century after the first suspicions by scientists. Even today, many people still smoke.

The biggest single problem, but certainly not the only problem in human-caused climate change, is carbon dioxide. We produce it when we burn oil and coal and natural gas to generate energy. Carbon dioxide traps heat in the atmosphere, adding to the natural greenhouse effect and causing climate change. A few farseeing scientists realized more than a century ago that this might happen. Yet accurate measurements of carbon dioxide amounts in the atmosphere began only in the late 1950s. Thus, we have known for only about half a century that the amount of atmospheric carbon dioxide is increasing. We ought to remember this half-century time scale when we get impatient about the slow pace of progress in action against human-caused climate change.

Incidentally, like many climate scientists, I do not fully approve of the catchy term "global warming," although I realize it is in the language to stay. It is an oversimplification. Climate is not just temperature. Climate is a rich tapestry of interlinked phenomena, multifaceted and inherently complex. The important aspects of climate change are local, not global, and are not confined to warming. Global warming is just a symptom of planetary ill health, like a fever.

You and your physician both know that fever is important but not the whole story. At your annual checkup, you do not confine yourself to body temperature when discussing your health. Even the most ignorant patient realizes that measuring temperature alone does not enable the physician to diagnose an illness and prescribe treatment.

Instead, everybody knows that a body temperature only a few degrees above normal is a symptom that can indicate health problems that may have serious consequences, including death. Yet we still have not educated most Americans to understand that a planetary fever of a few degrees can mean melting ice sheets, rising sea level, massive disruptions in water supply in the arid American west, increased risk of wildfires, killer heat waves, and stronger hurricanes on the Atlantic and Gulf coasts of the United States.

What can we say about hurricanes and their possible connection to an altered climate? The short answer is that you have to think about probabilities when you think about this connection. A warmer climate means that the strongest hurricanes may become even stronger, on average. It does not mean we can definitely prove that any particular hurricane owes its strength to climate change, only that the odds of very strong hurricanes have gone up.

A hurricane in simple terms is essentially just a heat engine for which sea surface temperature is an approximate indicator of the fuel supply. The higher the temperature of the ocean surface, the more energy is available to power the hurricane. There is a critical sea surface temperature of about 80 degrees Fahrenheit (about 26.7 degrees Celsius), below which hurricanes generally do not form. Because their destructive power increases as the sea surface temperature does, and especially because of the large recent increase in population and development in hurricane-prone areas in the United States, our vulnerability to hurricanes has increased strongly.

Scientists are cautious people, skeptical to a fault, fond of caveats, and not given to sweeping statements. We prefer to make claims only when we can back them up with solid data. We know that hurricanes are highly variable, no two are alike, and next year's hurricane season might be very different from this year's. It is our natural inclination to wait a few more years, observe more hurricanes, improve our theories and models, until we have an airtight case to present.

Nevertheless, the best current research tells us that the oceans have recently warmed substantially, that human activities are the primary cause of that warming, that an increase in the intensity of strong hurricanes is the expected result, and that we have indeed observed an increase in the numbers of the strongest hurricanes. No amount of waffling over probabilities and statistics can obscure these sobering results. Many intelligent people still laugh at the small numbers we use and think a global warming of a few degrees is trivial. They may say that moving from a colder city to a warmer one involves a much greater warming and is actually quite pleasant. These people just do not grasp the crucial difference between local changes and global ones. They do not realize that when the climate of the entire planet changes by a few degrees, enormous changes happen. Going into an ice age, to pick one example, involves a global cooling of only a few degrees.

Some people really think that a rapidly warming climate is just a minor inconvenience that can be handled by air conditioning and other minor technological fixes. This massive degree of misunderstanding may be due in part to a failure to educate people about science. It may also be the case that people have become confused by the widespread misperception that the science of climate change is immature, uncertain, characterized by raging controversy, and not to be trusted. An effective campaign of deliberate disinformation about climate science has helped spread this false impression.

Medical science has achieved a measure of pervasive respect that climate science can only envy. Journalists covering a medical discovery do not mistrust researchers and do not inevitably insist on hearing from "the opposing view." When reporting on research showing the need for Americans to eat more sensibly and be physically active, the media does not treat these advances in medical science in terms of a dispute. Journalists do not feel obliged to seek out medical contrarians "for balance."

On the other hand, perhaps the COVID-19 pandemic that began in 2020 can serve as a warning example. As with climate science, when the medical profession presented bad news, quite a few people turned against medicine. Prominent medical experts who appeared often on television and gave widely publicized advice found themselves transformed from heroes to villains. Also, conspiracy theories about vaccines flourished. Thus, medical science may illustrate why climate science sometimes is opposed rather than accepted. It may be the propensity of some people to turn against science when they do not like what it tells them.

There are many parallels between the climate change issue and medical topics. Perhaps some can be useful in educating people and politicians. It has turned out to be frustratingly difficult to get people and their governments motivated to act to avert climate change. Yet people are intensely interested in threats to their own health. Many Americans have improved their health by making major changes in their personal lives, changes that are directly attributable to the results of medical science. Real progress has been made in making Americans, and their government, more aware of unhealthy behavior. The media, including public service advertising, together with organizations such as the American Medical Association and the American Cancer Society, have succeeded in raising many people's consciousness about health.

In climate change, the comparable scientific organizations have made very little progress in persuading people. In fact, most of the professional societies that scientists like me belong to exist mainly to serve the scientific community. They organize conferences of researchers. They publish highly technical journals that only scientists can read. These societies have low profiles and are essentially invisible to the public. Most of these societies have tiny budgets and devote very little effort to outreach of any kind. Some appear to be politically inactive or naive.

It is also true that some powerful segments of industry vigorously oppose efforts to act and to publicize the scientific facts about climate change. However, business and industry are not monolithic in this respect. There are outstanding corporate champions of sound climate science, and we know that even the most retrograde segments of industry can change and become forces for progress, as notably happened in the ozone issue, for example. There, after it was scientifically proven that human-made chemicals were the culprit that caused the ozone hole, the industry that manufactured these chemicals changed its tune and developed safe substitutes for them. Governments and science and businesses cooperated, and humanity benefited.

In other cases, science and public concern have eventually triumphed over misguided opposition and propaganda. Numbers of smokers and deaths from smoking have been significantly reduced. Most Americans realize that smoking is dangerous and kills many thousands of people every year. They have learned this despite a highly professional and well-funded disinformation campaign mounted by portions of the tobacco industry.

Quitting smoking, like quitting using fossil fuels, is not easy to do, and in both cases the difficulty in quitting is immediate, while the most important benefits are all long-term.

The widespread public concern about the health consequences of smoking tobacco has led to political action, including warning labels on cigarettes, restrictions on advertising, and bans on sales to minors. The tobacco industry has repeatedly been defeated in court cases and has already paid large amounts of money as a result.

We see too the results of governments responding to public concern in the arena of promoting healthier food choices, including laws mandating truth in labeling and other actions to increase public awareness. These examples, and many more that could be cited, are direct results of medical science affecting public policy. People are persuaded that the science is right, and governments react to concern and pressure from citizens.

Science seems mysterious to many people, and it is not easy to penetrate the barriers of jargon and mathematics to explain the intricacies of computerized climate models or satellite climate measurements to a lay audience. Although very few people have a deep understanding of science or indeed any detailed familiarity with what researchers actually do, the public generally respects scientists and has confidence in the validity of their results. In fact, polls consistently show that scientists are among the most widely admired people in our society.

Risk is an inevitable aspect of life. Medicine involves risk. People tend to be realistic about the consequences of serious medical problems. They know that a bypass operation is major surgery. They accept the cost and the risk, understanding clearly that doing nothing also entails real costs and dangerous risks. They do not expect that a simple bandage will cure a potentially fatal disease. As a climate scientist, I sometimes fear that we are wasting time arguing about which type of bandage is most attractive as a climate remedy, instead of facing the hard decisions, and the risks, that climate change demands of us.

You cannot fool Mother Nature. The climate system responds to changes in the amounts of heat-trapping gases and other substances in the atmosphere. The climate system is indifferent to economic concerns, political considerations, or societal implications. The climate system does not care about the details of cap-and-trade agreements, and it knows nothing about diplomatic niceties such as protocols and framework conventions. The amount of carbon dioxide in the atmosphere is what matters most to climate.

The laws of atmospheric physics, unlike government reports, are absolutely immune from political tampering. If humanity insists on adding heat-trapping gases to the atmosphere, there will be consequences. That is just a fact. We scientists are busy researching the quantitative details, but we already know the big picture pretty well. If you see that a glib climate contrarian is not at all worried about doubling the amount of carbon dioxide in the Earth's atmosphere, then start to think about tripling, quadrupling, and beyond. That is where we are headed, and our speed on this wrong road is actually still increasing. To have an effect, we simply must do more than make small token reductions in greenhouse gas emissions.

In the Preface, I mentioned that Revelle and Suess (1957) wrote that humanity is doing an inadvertent and unrepeatable geophysical experiment in moving so much carbon to the atmosphere so quickly. That perception, visionary at the time, seems obvious now. What is still not obvious to many people is that all of us are now engaged in a second global experiment, this time an educational and geopolitical one. We are going to find out whether humanity is going to take climate science seriously enough to act meaningfully, rather than just waiting around until nature ultimately demonstrates that our climate model predictions were right.

In the end, our success or lack of it will be measured by whether we as a global society can change the Keeling curve (Figure 14.1), bending it downwards, and whether we can stabilize the amount of carbon dioxide in our atmosphere in time to avoid the most dangerous climatic consequences. Whether that will turn out to be possible is not yet known. I hope so. I think it is the single most important question in planetary public health: armed with impeccable science, can humankind muster the wisdom and the will to make difficult changes? With many medical issues, the outcome is ultimately in the hands of the patient. In this case, it depends on all of humanity.

The biggest unknown about future climate is human behavior. Everything depends on what people and their governments do. For centuries, we humans were passive spectators at the global climate change pageant. Not any longer! We have become the dominant actors. You and I, and all 8.2 billion people who are alive today (early in 2025), do indeed have our hands on the thermostat that will control the climate of our children and grandchildren. "The thermostat" (Figure 7.2) is a powerful metaphor. It is very useful in climate change science communication. People know that a household thermostat enables a person to control the temperature of the interior of a building. Yet many people do not realize that human activities are now a dominant factor in controlling climate change. We humans have caused the world to warm in recent decades. We human beings now are challenged to find the best ways to limit the warming, to turn down the thermostat, and to avoid some of the most disruptive consequences of severe climate change. This is a pressing problem, and it is one that arrives when humanity also faces other pressing problems. The big question is whether we together can meet this challenge successfully and are able to act promptly, wisely, and forcefully. Everything depends on what people and their governments do.

11
Language

Language is a critical aspect of all communication, written or oral, on every topic. When the topic is science, especially climate change science, some aspects of language become especially important. My first bit of language advice to anyone interested in communicating effectively about climate change science is to avoid the mistakes that climate scientists themselves often make. If you read enough research articles written by scientists, you are certain to discover that they tend to follow a peculiar format, one that is quasi-chronological. They typically start with background information, such as summarizing what previous research has been done on the particular topic being investigated. After that, the usual research paper moves on to an account of all the preparations that were made to start the research project being reported. Perhaps scientific instruments had to be procured or constructed and then tested. Perhaps an expedition to a remote location had to be arranged. Perhaps computer programs had to be written and revised. Measurements or observations may have been taken, then processed and analyzed. At the end of this very long story, the authors present their results and conclusions.

Having read thousands of scientific research articles myself, I sometimes think that scientists force themselves to follow this structure, even though the way the research was actually carried out may have been very different. I think it is helpful to realize that making a scientific discovery and finding new knowledge is like climbing a treacherous mountain trail, with no flashlight, on a pitch-dark night. Unable to see, the climber stumbles and falls often, suffers many severe bruises from rocks and many sharp cuts from thorny plants. Then the climber is forced to turn back several times and try alternative routes. Finally, the climber manages to reach the top of the mountain, despite being exhausted and in pain.

Only then, as the Sun comes up, is the climber able to see a smooth and gently sloping path, the best possible route to success, the route that should have been taken, a path that leads easily and painlessly from the bottom of the mountain to the top. When writing the article reporting the research, it is that smooth path that the scientist describes, starting with a calm and thorough search of previous work, then the making of orderly preparations for the research, and finally the logical carrying out of a swift and successful project, leading triumphantly to important new results. There is not a word about the long nightmare on the mountain and the many cuts and bruises endured on the perilous trail actually taken.

Do not follow the example of the scientist who communicates all the details and background first and then announces the results and conclusions at the end. In journalism, this sin is called "burying the lead." Reporters learn to compose a lead, the first sentence of a news story, in a way that conveys the main point of the story and also captures the reader's attention and motivates the reader to continue reading. We are speaking here of written stories and articles, but the point applies to oral presentations of all kinds too, including informal conversations. If a communicator has the scientist's unfortunate habit of giving a lot of unimportant details first, and the important main result last, I say, "Turn your world upside down." Start with the most significant point. Everyone should take this admonition seriously: "Do not bury the lead."

Scientists in any specialty tend to speak to one another in a strange and private language that seems bizarre to nonscientists or even to scientists in other specialties. Jargon and mathematical terms are part of normal conversational usage for scientists. Words that are unfamiliar to the wider world should be avoided. They always have clear and simple substitutes. Rather than "anthropogenic," scientists could and should say "human caused."

Be sure to use units that are familiar to your audience. Scientists everywhere use metric units in their work, and they often publish articles using these units. When speaking to or writing for a nonscientific audience in the United States, remember that metric units will be both unintelligible and frustrating to the audience. Instead, use feet and miles rather than meters and kilometers, use pounds instead of kilograms, and use degrees Fahrenheit rather than Celsius.

Scientific jargon refers to a type of language used by scientists in communicating efficiently and precisely with one another. Most scientists realize that jargon will not be understood by the public. However, there is an insidious trap involving common everyday terms that are not jargon, and that many people use, but that scientists use to mean something completely different from what everybody else means. There are hundreds of such terms, and they should be learned and avoided by anyone wishing to communicate climate change science effectively.

For instance, many climate scientists are shocked to learn that people misinterpret the term "positive feedback," which scientists always use to mean a self-amplifying process. Here is an example: a warming Arctic causes less snow and ice, and so it makes the surface of the Earth darker. That darker surface is then less reflective, so it absorbs more sunlight, increasing the warming. This process is one of the main reasons why Alaska and other locations in the Arctic have warmed much more in recent years than the global average. For scientists, such a "positive feedback" increases global warming or climate change and thus is clearly bad. These scientists have forgotten that it is normal for people to be delighted when their boss praises their work, thus giving them "positive feedback"!

Similarly, when human activities add heat-trapping gases to the atmosphere, climate scientists frequently refer to the consequence as an "enhanced greenhouse effect." They mean that the natural greenhouse effect is increased in strength, producing more warming in the atmosphere and causing climate change. In using "enhanced" to mean "intensified," these scientists overlook the everyday meaning of "enhanced," which is "improved," as when attractive clothing or good health is said to enhance a person's appearance. Thus, just as in the case of "positive feedback," scientists intend to describe something harmful and undesirable with the word "enhanced," but their use of the term is confusing and creates misunderstanding, because normally "enhanced" describes something beneficial and desirable. I urge every climate change science communicator (including you when you speak to your family and friends and colleagues about climate change) to heed this warning. Try to compose messages that are simple and memorable, to repeat them often, and to partner with trusted messengers. I heartily endorse the use of metaphors and other vivid imagery. If climate change is very important to you, do not speak or write about it in dry and unemotional language that conveys boredom and resignation. Instead, let your passion show. Seize opportunities to learn from expert communicators and to get useful feedback from your audiences. Nobody is born knowing how to ski or play chess or drive a car. Like all these skills, communication skills can be taught, developed, practiced, and improved.

12
Solutions

Science is based on facts and evidence. Science can inform decisions about solutions, but it cannot provide 100% of all the relevant information. As a climate scientist, I realize that I do not know all the relevant facts and evidence about solutions. However, as one citizen of the planet, I may have opinions. In this chapter, I include several of my opinions about solutions.

Solutions are vitally important. When you communicate climate change science to anyone, be sure to include information on solutions. Nobody wants to hear about hopelessness, and in the case of climate change, there are many reasons to be hopeful. Climate change poses difficult problems and challenges, but there are lots of aspects of solutions that are both creative and practical and that can help solve the problems and overcome the challenges of climate change.

In discussing choices and solutions, I will say at once that my own expertise is in atmospheric and climate science. These are the areas of my research and teaching experience. My university degrees are in meteorology. I am definitely not an expert on energy systems, for example, or on economics or politics. It would be presumptuous of me to act as if I had all the answers and was certain about my knowledge of the best solutions. For example, using nuclear power to generate electricity is an alternative to generating electricity from burning fossil fuels. Nuclear reactors do not emit carbon dioxide. Thus, relying on nuclear power is one route to decarbonizing the world's energy system.

I am unqualified to lecture anyone about the strong and weak points of nuclear power. In my opinion, there are historically four major objections to nuclear power. They are cost, reactor safety, proliferation, and waste. In my opinion, there has been important recent progress in all these areas. In my opinion, the urgency with which the world needs to quickly and drastically reduce emissions of carbon dioxide means we should not rule out nuclear power. In my opinion, it needs to be part of the mix. It is already a significant component of the mix for both the United States and the world. I have lived in France, where about two-thirds of the electricity is generated using nuclear reactors. I know that polling data consistently show that most citizens of France approve of nuclear power. They also appreciate that nuclear power helps to make France able to independently meet its own energy needs, without being obliged to purchase fossil fuels from other countries. Globally, about 10% of our electricity comes from nuclear power. I favor objective and unbiased consideration of nuclear power, and I disapprove of opposition to nuclear power when it is based only on emotional or ideological reasons. In my opinion, it would be wise to learn from the experience of countries that have chosen to rely on nuclear power.

In the area of solutions to the challenges of climate change, I also think the main barriers to action today are not all technical or financial. The barriers also include a lack of wise and inspiring political leadership. Science can help to inform policy, but only concerned people and responsive, capable governments can decide what policies are best and then implement them.

In the United States today, for example, we clearly do not yet have national agreement on climate change. Despite the strong scientific consensus, climate change is controversial politically. Incidentally, there are many lawyers in Congress, but very few members of Congress are scientists who hold PhD degrees in science. Perhaps that should change!

There are many approaches to reducing dependence on fossil fuels, including increased energy efficiency and energy conservation and much more use of sun, wind, and water to provide the energy the world needs. These renewable resources are widely available now and can sometimes appear to be cost-competitive with fossil fuels. However, most of the available sites for hydroelectric power generation are already in use. Also, both solar power and wind power suffer from intermittency. The electricity they produce is not available all the time. Thus, relying on these renewable resources involves providing means for storing electric energy, so that it will be available at times when it is not being generated. One colleague whose expertise I respect tells me, "We are getting to 40 per cent or 50 per cent renewables reasonably rapidly, but going beyond that will be very expensive unless there is a breakthrough in energy storage technology."

The nations of the world agreed at the COP 21 meeting in Paris in December 2015 to limit warming to 2°C, or 3.6°F, above preindustrial levels. At the urging of the most vulnerable countries, the delegates in Paris also agreed to "pursue efforts to limit the temperature increase even further to 1.5 degrees Celsius," although it was widely understood that successfully meeting this aspirational goal would be much more difficult than meeting the 2.0°C target.

After Paris, is the glass half empty or half full? I am guardedly optimistic. I see many reasons for optimism, and I encourage everyone to learn about them and communicate them to their own Uncle Pete and to their other audiences:

- World leaders are engaged, at least almost all of them are.
- Emissions of heat-trapping gases have begun to decline in some countries.
- Solar and wind energy have dropped drastically in price and continue to become cheaper.
- Renewable energy use in many countries is increasing rapidly.
- Many corporations are now acting to reduce emissions.
- States and localities are acting too.
- Many countries are showing rapid progress.

I am also encouraged by recent polling that shows that in the United States, more people accept the science and are very concerned about global warming than was the case only a few years ago. However, as long as virtually the entire leadership of the Republican party in the United States rejects the findings of mainstream climate change science and considers climate change to be a hoax, we clearly will have a long way to go. In the United States, climate change has become a very partisan issue. However, I am guardedly optimistic. Market forces now favor carbon-free energy. Coal companies are going bankrupt. Renewable energy gets cheaper every year. Recent developments in nuclear power are encouraging. Electric vehicles are happening quickly. Much energy policy in the United States is set at state and local levels, not by the national government.

Free-market, small-government mechanisms, such as revenue-neutral carbon fee-and-rebate plans, can work. Some have been advocated by leading conservatives, who argue that it is sensible insurance, hedging against climate change risk, whether one accepts climate science or not. Leaving a healthy climate to your children and their descendants is a worthwhile goal, and a realistic one. It will not be easy, and it will not happen as quickly as we would like, but it can be done.

We need to help people realize that not acting is also making a choice, one that commits future generations to serious climate change impacts. Research suggests that messages that may invoke fear or dismay are better received if they also include hopeful messages. Thus, we can improve the chances that the public will hear and accept the science, if we include positive messages about our ability to solve the problem. For example, we can explain that future climate is in our hands; lower emissions of heat-trapping gases will mean reduced climate change and less severe impacts. We can point out that addressing climate change wisely can yield a variety of benefits to the economy and quality of life. We can explain that acting sooner is preferable to delaying. We can all rise to the challenge of helping the public understand that science can illuminate the choices that we face.

Whether to act to limit global warming to tolerable levels should not depend on your politics. We have only one Earth. Everybody should want to avoid polluting and contaminating this magnificent world, and everybody should agree that we need to protect and preserve our amazing planet. Your policies and values and politics have a role to play in deciding which actions are best, but any rational policy begins by accepting the science.

The world needs to take firm action about the threat of human-caused climate change within the next decade. Research shows that global emissions of heat-trapping gases must peak and decline quickly – within a few years, not a few decades or centuries – if global warming is to be limited to a level that avoids severe climate disruption. Meanwhile, a well-funded and effective professional disinformation campaign has been successful in sowing confusion, and many people mistakenly think climate change science is unreliable or is controversial within the scientific expert community. Thus, an urgent task for us scientists and for all communicators of climate change science may well be to give the public guidelines for recognizing and rejecting junk science and disinformation. If very young students today, who will be adults tomorrow, can understand and apply these guidelines, they may not need to acquire a detailed knowledge of climate change science. To that end, I offer the following six principles.

1. The essential findings of mainstream climate change science are firm. The world is warming. There are many kinds of evidence: air temperatures, ocean temperatures, melting ice, rising sea levels, and much more. Human activities are the main cause. The warming is not natural. It is not due to the sun, for example. We know this because we can measure the effect of human-made carbon dioxide, and it is much stronger than that of changes in the sun, which we also measure. The greenhouse effect is well understood. It is as real as gravity. The foundations of the science are more than 150 years old. Carbon dioxide in the atmosphere traps heat. We know that, because careful laboratory experiments prove it, and theoretical physics explains it. We know carbon dioxide is increasing, because we measure it. We know the

increase is due to human activities like burning fossil fuels, because we can analyze the chemical evidence for that.
2. Our climate predictions are coming true. Many observed climate changes, like rising sea level, are occurring at the high end of the predicted range. Some observed changes, like melting sea ice, are happening faster than some recently anticipated worst cases. Unless humankind takes strong steps to halt and reverse the rapid global increase of fossil fuel use and the other activities that cause climate change, and does so in a very few years, severe climate change is inevitable. Urgent action is needed if global warming is to be limited to moderate levels.
3. The most familiar climate change myths and falsehoods have been refuted many times over. The refutations are on many websites and in many books. For example, the mechanisms causing natural climate change like ice ages are irrelevant to the current warming. We know why ice ages come and go. That is due to changes in the Earth's orbit around the sun, changes that take thousands of years in order to affect the geographical distribution of sunlight on the Earth. The warming that is occurring now, over just a few decades, cannot possibly be caused by such slow-acting processes. However, it can be caused by human-made additions of heat-trapping substances to the atmosphere.
4. Science has its own high standards. Science does not mean unqualified people, who do not carry out scientific research, making unsubstantiated claims on television or the internet. Science means expert scientists doing research and publishing it in carefully reviewed research journals. Other scientists examine the research and repeat it and extend it. Valid results are confirmed, and wrong ones are exposed and abandoned. Science in the long run is self-correcting. People who are not experts, who are not trained and experienced in this field, who do not do research and publish it following standard scientific practice, are not doing science. When they claim that they are the real experts, they are not being truthful.
5. The leading scientific organizations of the world, such as national academies of science and professional societies of scientists in fields relevant to climate change, have carefully examined the results of climate science and endorsed these results. It is silly to imagine that thousands of climate scientists worldwide are engaged in a massive conspiracy to fool everybody. It is also silly to think that a few minor errors in the extensive IPCC reports can invalidate the reports. The first thing that the world needs to do to confront the challenge of climate change wisely is to learn about what science has discovered and accept it.

6. One last time: Always remember why we need to communicate climate change science. We need to inform people. We need to motivate them. We need them to act. The biggest unknown point about future climate is human behavior. The choice is up to us. Everything depends on what people and their governments do.

PART III

Scientific Investigations of the Climate System

Introduction to Part III

The fundamental facts about climate change are crucial and should be emphasized by anyone studying climate change. The media bring us news about climate change continually. News is new, and it often tends to be spectacular as well as novel. However, context may be missing. Today we may be told about a research group that thinks a large ice sheet in Antarctica may destabilize. Yesterday we may have heard about a record-breaking heat wave in Africa. Tomorrow we may learn of progress toward finding an efficient way to remove carbon dioxide from the air. All these developments are interesting, and some may turn out to be very important, but they should not be allowed to obscure the fundamental facts that do not change. We have known about the fundamental facts of climate change for a very long time. They are still valid. These truths are central and unchanging.

This part is largely based on Chapters 3 and 4 of my book, *The Forgiving Air: Understanding Environmental Change* (Somerville, 2008). The material has been condensed and revised and brought up to date in 2025. This book, now out of print, is based on my lectures given at the University of California, San Diego, to schoolteachers. The material from *The Forgiving Air: Understanding Environmental Change* is reproduced here with permission from the University of California Press and the American Meteorological Society. All rights to this book reverted to the author, Richard C. J. Somerville, in 2004.

13

The Greenhouse Effect

By now many people have at least a basic notion of what the greenhouse effect is all about. There are gases in the atmosphere – including water vapor, carbon dioxide (CO_2), methane, nitrous oxide, ozone (O_3), and chlorofluorocarbons (CFCs) – that act somewhat like the glass of a greenhouse. They are partially transparent to sunlight. As the sunlight passes through the Earth's atmosphere, some of the ultraviolet radiation in the sunlight gets absorbed by ozone, and some other portions of the sunlight get absorbed by other constituents of the atmosphere, including clouds, but much of the sunlight reaches the Earth's surface unimpeded by these gases in the atmosphere. This sunlight, or solar radiation, is largely absorbed at the surface of the Earth. This energy warms the surface and is then re-emitted as infrared radiation, or heat. But these same "greenhouse gases" that were largely transparent to the incoming sunlight are not transparent to the infrared radiation, or heat, that the Earth emits. They, together with clouds, absorb some of it, and part of what they absorb is radiated back toward the surface of the Earth. The overall effect of these gases is to trap some of the heat within the atmosphere. Thus, somewhat like a blanket on a bed, the greenhouse gases make the Earth warmer than it would otherwise be.

You might think that the science of climate change is young, and that scientists developed it only a few years ago. That is not true. One of the most important pioneers, who did superb research and played a major role in creating the science of climate and climate change, was born more than 200 years ago. John Tyndall (1820–1893) was one of the world's most eminent experimental physicists of his time. Born in Ireland, he worked mainly in England, eventually becoming a professor of natural philosophy in London at the Royal Institution of Great Britain and ultimately succeeding Michael Faraday as its superintendent.

Tyndall built on earlier theoretical and conceptual work by the French physicist and mathematician Jean-Baptiste Joseph Fourier (1768–1830) and others.

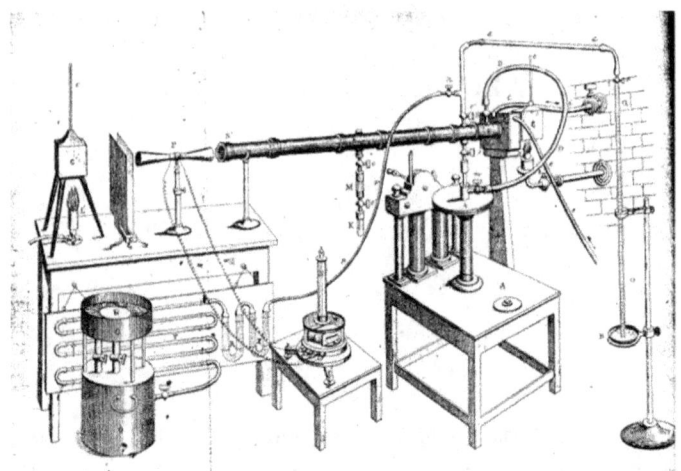

Figure 13.1 Tyndall's experimental apparatus, which enabled him to put the greenhouse effect on a firm empirical scientific foundation, shown in this drawing from his original publication describing this research (Tyndall, 1861).

In his pioneering laboratory research, Tyndall was the first to put the concept of the greenhouse effect on a firm empirical foundation. He accomplished what nobody else had done. Using equipment of his own design (Figure 13.1), he carefully measured the relative ability of water vapor, carbon dioxide, and other atmospheric gases to absorb infrared radiation. Tyndall immediately realized the potential implications of his discoveries for climate change. The year 2011 marked the 150th anniversary of the publication of his definitive paper reporting the results of this research (Tyndall, 1861).

In addition to his original research in experimental physics, John Tyndall was a superb lecturer and author who spent a significant amount of his time disseminating science to the general public. He wrote many popular books and articles and gave hundreds of public lectures. The income from these widely praised educational efforts made Tyndall a wealthy man, and he gave significant sums to support science. For those of us interested in climate change, the highlights of John Tyndall's scientific career are his contributions to atmospheric and climate science through both research and outreach.

John Tyndall was born on 2 August 1820 in Leighlinbridge, County Carlow, Ireland. His ancestors had come to Ireland from Gloucestershire in England in the seventeenth century. Tyndall's family was not wealthy, and his father was a shoemaker and a member of the local police force, the Irish Constabulary. His father played a major role in Tyndall's life and inspired in John a love of learning and debate. John remained in school until age 17, an unusually late age for a boy in those circumstances.

After working as a surveyor, he then became a science and mathematics teacher in an experimental Quaker school located near Stockbridge in Hampshire, England. The school, Queenwood College, was highly innovative and was dedicated to science teaching. With a friend, Edward Frankland, Tyndall both studied and taught. Aware of their own limited expertise, they had decided by 1848 to go to Germany for additional training.

The two arrived in Marburg, Germany, in October 1848. Robert Bunsen (1811–1899), a professor at the University of Marburg and a distinguished chemist, made room for them in his laboratory. Bunsen (known to chemistry students today as the inventor of the Bunsen burner, a common piece of laboratory equipment) was already well known. He later discovered two chemical elements, cesium and rubidium. Tyndall at that time had only a limited knowledge of mathematics, science, and German. However, he was determined and worked intensely, and Bunsen was known to be an inspiring teacher. By 1850, Tyndall had completed his doctorate.

He was warmly welcomed by the German scientific community. Tyndall arrived back in England in 1851. He sought work. To earn a living, he returned to the school at Queenwood. In 1852, he was elected to the Royal Society, the United Kingdom's national academy of sciences. However, Tyndall received little recognition in Ireland, the country of his birth, during his lifetime.

The decisive turning point in Tyndall's career came on 11 February 1853. He gave a brilliant lecture that day at the Royal Institution of Great Britain in London. The Royal Institution is an organization for scientific education and research, founded in 1799. Michael Faraday, Professor and Director at the Royal Institution and a scientific superstar, was in the audience and was very impressed. Tyndall gave additional lectures and was offered a position in May 1853 at the Royal Institution. There he was elected Professor of Natural Philosophy. Other job offers arrived too, but the opportunity to work with Faraday at the Royal Institution was decisive.

Tyndall would remain at the Royal Institution for 34 years. It offered him an ideal opportunity to do experimental work. It was also an excellent stage for Tyndall's talents as a speaker and science popularizer. As a lecturer on science to the public, Tyndall was magnificent. He was obsessive in planning and preparing his lectures. He devised and rehearsed spectacular demonstrations. He had a superb sense of showmanship and great style. He ranked with Faraday and Huxley as a popular speaker. For comparison, perhaps Carl Sagan or Jacques Cousteau in the recent past, or Neil deGrasse Tyson today, might be Tyndall's modern equals as scientists who also became charismatic and well-known public speakers. Tyndall's lectures were the basis of many articles and books, and he ultimately became quite well known.

In 1876, John Tyndall married. He was then 56 years old. His marriage, to Louisa Hamilton, was a happy one. They had no children, but she shared his many interests. These included Tyndall's love of Alpine adventures. Together, they built a home in the Swiss mountains. Ill health caused Tyndall's resignation from the Royal Institution in 1887. By then, his scientific career had ended. During an illness, Tyndall died tragically from an accidental drug overdose, administered by his wife, on 4 December 1893.

Tyndall's research interests were very broad. Climate science today values his work on gases and radiant heat. He also did first-rate research on glacier motion, diffusion of light in the atmosphere, the germ theory of disease, and many other topics. Tyndall received many scientific honors and succeeded Faraday as the Superintendent of the Royal Institution (Jackson, 2018).

It must be said that Tyndall enjoyed a good intellectual fight. He has been described as "a keen controversialist" and as "the very model of an Irishman: wild, athletic, a hard worker and a fluent talker." He had strong political views and was likely an agnostic. His support of Darwin's theory of evolution brought him into conflict with the church. He was also a unionist, favoring continued political union between Ireland and Great Britain. Tyndall was personally generous. He gave large sums of money to support young American scientists who followed his example of going to Germany to study science. From the mid-1800s until perhaps the 1930s when the Nazis came to power, Germany was a leading center in mathematics and the natural sciences, such as chemistry and physics. Germany produced numerous world-class scientists, and German universities attracted students such as Tyndall, who sought an outstanding scientific education.

Tyndall's research on the absorption of "radiant heat" or infrared radiation by gases is of primary importance to climate science. Fourier and others had theorized and speculated about it. In fact, infrared energy had been discovered only in 1800, by William Herschel, the German-British astronomer and composer of music. Infrared energy was still largely mysterious in mid nineteenth century. Tyndall explored what was later dubbed the "greenhouse effect" empirically. He invented the necessary laboratory equipment, and he made the key measurements himself.

In the paper reporting the results of these experiments (Tyndall, 1861), it is clear that Tyndall had immediately realized the significance of his discovery for climate. He wrote that "a slight change" in the atmospheric amount of carbon dioxide or other infrared absorbing gases could have important effects on climate: "Such changes in fact may have produced all the mutations of climate which the researches of geologists reveal."

After Tyndall, others investigated the effect of changes in atmospheric carbon dioxide on the climate. In the 1890s, Svante Arrhenius, a renowned Swedish physicist and chemist (Crawford, 1996) who would later receive the 1903 Nobel Prize in chemistry, did the first serious scientific calculations of the sensitivity of the climate to carbon dioxide (Arrhenius, 1896). He estimated an equilibrium climate sensitivity of about 6 degrees Celsius warming (10.8 degrees Fahrenheit) that would result from a doubling of the amount of CO_2 in the atmosphere. In a book published in 1906 and in English two years later (Arrhenius, 1908), he decreased that estimate to 4 degrees Celsius (7.2 degrees Fahrenheit). This number is within the range of modern estimates. Arrhenius was both brilliant and lucky to be so accurate. He was deeply insightful in understanding the relevant physical processes. For example, he realized that warming caused by additional carbon dioxide would result in increased water vapor in the atmosphere, which would in turn amplify the warming. He also had a deep understanding of the relevant theoretical physics of that time. His calculations, all done by hand in the precomputer era, were extensive and accurate. In that sense he was brilliant. But the science of the day was too crude, and Arrhenius made quantitative errors in both spectroscopy and atmospheric modeling, errors that fortuitously nearly canceled one another out. In that sense, Arrhenius was simply lucky.

Arrhenius did this research because he was interested in the natural causes of ice ages, not in the possibility of human-caused climate change that motivates modern research in climate change. Today, we understand clearly that the climate is indeed sensitive to carbon dioxide. Advances in theory and observations have led to a robust science of climate change. Human-caused increases in atmospheric carbon dioxide have already produced observed climate change, and much more serious climate change will occur in the future if emissions of carbon dioxide continue to increase.

It can be useful to imagine an alternative history of our planet, one where humankind exploited fossil fuels for generating energy, just as has actually occurred, but did not develop satellites, supercomputers, climate models, and the other essential tools of modern climate science. Let us call the imaginary planet with that alternative history "Planet B." The climate of Planet B would be changing, due to human activities, just as the climate of the actual Earth is now changing, but scientists on Planet B would not be able to observe, understand, or predict this change.

Today, people who reject climate change science are effectively living on Planet B. An efficient and well-funded disinformation campaign promotes the denial of climate science. Its goal is to defeat policies that would limit greenhouse gas emissions. This disinformation effort is partially responsible for the fact that many nations throughout the world have largely failed to reduce emissions. Instead, the global amount of emissions of carbon dioxide has actually increased. The details of how this climate science disinformation campaign works and how it resembles other efforts to cast doubt on the findings of scientific research are the subjects of a remarkable book by two historians of science: *Merchants of Doubt* (Oreskes and Conway, 2010).

There are many possible reasons for the world's failure to act more quickly and more strongly to limit human-caused climate change. Among them, there is surely a great need today for more effective understanding by more people of the results of modern climate change science. The objective of this book is to help us all reach that goal.

Tyndall's laboratory equipment is depicted in a line drawing (Figure 13.1), which first appeared in the scientific paper he published describing his experiments that showed which atmospheric gases absorbed infrared radiation and thus were key to the Earth's greenhouse effect (Tyndall, 1861). I had seen this line drawing many times in studying Tyndall's research. I think I must have assumed that everything in Tyndall's laboratory had disappeared long ago. Thus, I was surprised and delighted one day when I had the opportunity to see and handle Tyndall's actual equipment.

Figure 13.2 is a photograph of a part of Tyndall's laboratory equipment. I made this photograph at a conference in Dublin, Ireland, in 2011. The

Figure 13.2 Several items of John Tyndall's laboratory equipment, with which he discovered that carbon dioxide and water vapor strongly absorb infrared radiative energy, while oxygen and nitrogen do not. Thus, he found the causes of the greenhouse effect. Photo credit: Richard C. J. Somerville.

conference was held to celebrate the 150th anniversary of the 1861 publication of Tyndall's scientific paper describing his remarkable discoveries. It was a moving and emotional experience for me to handle the original equipment that Tyndall had designed and constructed and to read the original notebooks in which he had recorded the results of his experiments. These articles were loaned to the conference by the Royal Institution of Great Britain, where Tyndall had worked. Two young women who were curators from the Royal Institution brought the articles to the conference in Dublin. I asked them how they had found these remarkable items, which were at least 150 years old. They replied, "It was easy. Every time a new Director is appointed at the Royal Institution, in order for the new Director to have a laboratory with his own equipment, the staff of the Royal Institution stores the previous Director's equipment in a closet. We just looked in the closet."

Except for water vapor, greenhouse gases are trace gases, meaning that they are present in very small amounts, or concentrations, typically just a few parts per million or even less. The most abundant gases in the atmosphere, nitrogen and oxygen, combine to make up about 99% of the molecules in dry air, but, as Tyndall discovered, they contribute nothing to the greenhouse effect. When the Earth radiates heat in the form of infrared radiation out toward space, it is the greenhouse gases – water vapor and the trace gases – that trap some of it. Clouds, too, are big contributors to the greenhouse effect, because they also trap heat. When you are outdoors on a cloudy night, you may notice that clouds keep you warmer than you would be on a similar but clear night.

Most sunlight reaching the Earth passes through the atmosphere unimpeded. The Earth is warmed mainly by the sunlight that is absorbed at the surface, both the land and the sea. Some sunlight, about 30%, is reflected back to space, mainly by clouds and thus does not interact with the climate system at all. This fraction of solar radiation that is reflected is called the albedo of the Earth.

The Earth radiates heat in the form of infrared energy; it does not radiate visible light as the Sun does. This occurs because the Sun is much hotter than the Earth, and the science of physics teaches us that, because it is so hot, the Sun must radiate at much shorter wavelengths than does the Earth. It is this infrared energy radiated outward from the Earth toward space that is partially trapped by the greenhouse gases in the atmosphere.

Real greenhouses, by the way, do not work entirely by the greenhouse effect. One of the ways a greenhouse keeps plants warm is simply by keeping wind out. When wind blows across plants, it carries heat away. The glass in the greenhouse, aside from whatever it does to the radiation, helps keep plants warm just by reducing the circulation of the air in the greenhouse.

There are many convincing demonstrations of the greenhouse effect in the atmosphere. A clear example of the impact it can have on the climate is seen when comparing the Earth and the Moon. Both the Earth and the Moon are about the same distance from the Sun, and both get their energy from the Sun. However, the Moon's albedo, or reflectivity, is lower than the Earth's, so you might expect that its climate would be warmer. In reality, however, the Moon is much colder than the Earth. The average temperature difference between them is about 35 degrees Celsius (63 degrees Fahrenheit), and this difference is due almost entirely to the presence or absence of a greenhouse effect. That is, the Earth has an atmosphere that traps heat, but the Moon has no atmosphere, and thus no greenhouse effect, which leads to a brutal day/night temperature difference, with scorching hot days and bitter cold nights. This is because there is no atmosphere to mediate the temperature change between day and night and make sunrise and sunset less abrupt.

The Moon is so small, and its gravity is therefore so weak, that any atmosphere it once may have had has long ago escaped to space. What surrounds the Moon, all the way down to its surface, is empty space. The early Moon may have had an atmosphere, but small planets, and small moons of planets, no longer have atmospheres because they are not gravitationally strong enough to retain them. This lack of an atmosphere occurs for the same reason that the astronauts frolicking on the Moon could make gigantic leaps almost effortlessly. The Moon has very little gravity. The Earth has sufficient gravity to have retained an atmosphere. The giant planets such as Jupiter and Saturn, with their massive gravitational fields, have very heavy, very dense atmospheres.

The greenhouse effect can also be seen on other planets. Compare the climates of Mars and Venus. Both have mostly carbon dioxide in their atmosphere, but Venus has a very dense atmosphere, about 100 times denser than Earth's. On the other hand, the Martian atmosphere is only about 1% as dense as Earth's. As a result, the greenhouse effect is very weak on Mars and very strong on Venus. Mars is very cold, and Venus is very hot and very cloudy. In fact, you need radar and a satellite just to look at Venus, for two reasons. The first is that you cannot even "see" the surface unless you use the special capability of radar to penetrate the cloud cover. The second is that you would have difficulties placing an instrument on the surface of Venus. It has been done, using space probes sent from the Earth, but the instrument does not survive long, because the surface of Venus is hot enough to melt metal.

This might be a good point to mention a historically important aspect of our understanding of the greenhouse effect. In 1856, a brief paper entitled "Circumstances affecting the heat of the Sun's rays" was published in a scientific journal called *The American Journal of Science and Arts*. The author was Eunice Newton Foote (1819–1888). She is not well known, but she was one of the first female scientists in the United States. The paper reported that Foote had carried out experiments in which she filled glass tubes with different gases, including carbon dioxide and moist air. She then placed these tubes in sunlight and in shade. She measured the resulting temperatures. The tube filled with carbon dioxide warmed especially strongly when it was in sunlight.

Eunice Foote deserves great credit for calling attention to these results. She was the first scientist to discover and report that changing the atmospheric amount of these different gases might affect the climate. However, after her 1856 paper was rediscovered recently, a certain amount of nonsense has been written about her work. We know today that some solar radiation (also called sunlight or shortwave radiation) is indeed absorbed by the atmosphere and therefore does not arrive at the Earth's surface. However, this absorption of solar radiation by the atmosphere is not the main reason why adding carbon dioxide to the atmosphere warms the climate. In fact, absorbing some sunlight in the atmosphere might prevent that amount of sunlight from reaching and heating the Earth's surface. Instead, we now understand that carbon dioxide in the atmosphere warms the climate primarily because it absorbs some of the infrared (or longwave) radiation that the Earth emits. This is the well-known greenhouse effect. We should recognize and admire Eunice Foote for what she did, especially because she did it in an era when women were often unwelcomed and unrecognized in science. However, we should not say that Eunice Foote explained the greenhouse effect, when she actually did not do that.

Foote did not investigate the greenhouse effect or the ability of different gases to absorb infrared radiation. There is no way that she or anyone else could discern that information from her reported experimental results. From her description of her experiment, it appears that her data would clearly be influenced by several sources of energy in addition to absorption of direct solar radiation reaching the tubes. We do not even know whether Eunice Foote was aware of the discovery of infrared radiation by Herschel in 1800. In the mid nineteenth century, much about infrared radiation was not yet known. It was largely mysterious, much as dark matter is today.

John Tyndall is correctly credited with discovering that carbon dioxide and water vapor in the atmosphere are strong absorbers of infrared radiation and are thus responsible for the greenhouse effect. He also correctly stated that changes in the atmospheric concentrations of carbon dioxide or water vapor can influence the climate. There is no evidence that Tyndall knew of Eunice Foote's research or that he had seen her published paper. It would have been entirely out of character for Tyndall to know of her work and yet fail to acknowledge it. Tyndall was honest and honorable throughout his life. He was always scrupulous about recognizing the contributions of others.

As we have seen, Svante Arrhenius was the first scientist to calculate how much temperature would change in response to a given change in the amount of carbon dioxide. However, a full theoretical understanding of the ability of different gases to absorb infrared radiation was not possible before the development of quantum mechanics in the early twentieth century.

Eunice Foote was the first person to carry out climate-relevant scientific experiments on different gases and to understand that changes in the amount of these gases in the atmosphere could alter the temperature. She deserves great credit for her priority in these important discoveries, especially for her explicit realization that climate change could have resulted in the past, if the amount of carbon dioxide in the atmosphere had been larger in the past than it is at present.

The greenhouse effect is not new, and today it is not mysterious. It has been keeping the planet habitable for billions of years. Without the greenhouse effect, we now know that the average surface temperature of the Earth would be well below the freezing point of water.

The concern about the greenhouse effect does not involve whether it exists. It is as real as gravity, and we should be grateful for it. The worry is that we human beings are inadvertently strengthening the greenhouse effect in a variety of ways by adding more of these trace gases to the atmosphere. When we refer to the greenhouse effect as a problem, we are being somewhat careless with our terminology. What we really mean is that the human-caused, or anthropogenic, strengthening of the greenhouse effect is the problem. This is an unnatural change that we humans are inadvertently imposing on the planet by changing the chemical composition of the atmosphere.

We are adding carbon dioxide (CO_2), the main culprit in the greenhouse problem, to the atmosphere, mainly by burning coal, oil, and natural gas. These are the so-called fossil fuels. About 80% of today's global energy supply comes from burning fossil fuels, thus producing CO_2. Fossil fuels provide cheap and abundant energy that generates electricity and heat. They provide power not only to factories and homes but also to ships, cars, trucks, trains, and planes. As a result, all over the world, CO_2 comes out of chimneys and out of tailpipes. It accumulates in the atmosphere.

We are also adding methane to the atmosphere in a variety of ways. It is generated from cows, for example, and human activities are responsible for the large number of cows on the Earth. Methane is also emitted into the atmosphere from rice paddies, landfills, leaks in natural-gas pipelines, and other sources originating in human activities.

In the main, the greenhouse gases we produce are by-products of the modern industrial world. For example, there is no efficient gasoline-burning engine that does not produce CO_2. In fact, CO_2 is an inevitable by-product of combustion. It is in many ways an innocuous gas. The bubbles in soft drinks, beer, and champagne are made of CO_2. It is colorless, odorless, and tasteless. But it contributes to the greenhouse effect.

It may one day be possible to develop an effective and economical way to capture the carbon dioxide that is created from, say, a power plant generating electricity from fossil fuels, before it is released into the atmosphere. The captured carbon dioxide could then be stored in a safe place, preventing it from getting into the atmosphere and further increasing the greenhouse effect. The technical term for this sort of CO_2 capture and storage is sequestration. A great deal of research is now underway aimed at making sequestration practical and economical on a large scale. However, barring a breakthrough, it seems unlikely that sequestration will be a major contributor to the reduction of greenhouse gas emissions in the near future.

The scientific consensus is that humanity is already confronted with climate changes that are not natural. We are now seeing unambiguous, incontrovertible, impossible-to-deny climate changes arising from human activities. These include warmer temperatures, reduced ice and snow, rising sea levels, altered patterns of rainfall, and many other aspects of climate change as well.

Unfortunately, the nature of the problem is such that the longer humanity waits to react, the greater the problem will be. Scientists use the word "committed" in this context to refer to future change that will occur because of past human actions. The scientific consensus is that because of the CO_2 we have already put into the atmosphere up until this point, we are committed to a certain amount of climate change even if we were to suddenly stop adding CO_2, which we are unable to do. Furthermore, humanity has not yet discovered and agreed upon a practical and relatively simple "technological fix" for global warming, analogous to the development of ozone-safe substitutes for CFCs. Instead, there are competing paths to attaining energy sufficiency while reducing dependence on fossil fuels, and we therefore do not yet have global agreement on how best to go about reducing emissions of CO_2 and the other greenhouse gases and other heat-trapping substances.

Furthermore, although there is a scientific consensus on the issue of climate change owing to a strengthening of the greenhouse effect, it is by no means unanimous. Some scientists dissent from it. In reality, unanimity in science is exceedingly rare and is certainly not necessary for a science to advance and to be useful. Here, we will examine which parts of the scientific evidence seem firmly established, and which remain less certain pending the results of future research. Right now, we have the exciting experience of tracking a branch of scientific research as it evolves, in contrast to a subject like plane geometry that was intellectually completed long ago and now is found in textbooks.

The effect on the climate of a strengthened greenhouse effect started out, from the scientist's point of view, as simply an interesting abstract problem, an experiment. As mentioned in the Preface of this book, two scientists, thinking about atmospheric CO_2 amounts increasing, remarked in the 1950s that "human beings are now carrying out a large-scale geophysical experiment of a kind that could not have happened in the past nor be reproduced in the future" (Revelle and Suess, 1957). They did not think of the process as anything that would necessarily be catastrophic or calamitous. It was just an interesting phenomenon that scientists could observe and study, with the Earth as a laboratory.

We have described the research of Svante Arrhenius, who in the 1890s had quantitatively calculated the climatic consequences of an increased greenhouse effect. An eminent chemist and physicist, Arrhenius was from Sweden – a relatively cold location – and he looked at the idea of the planet warming up as something that, on the whole, would be beneficial. Arrhenius's final words on the greenhouse effect are not at all gloomy: "By the influence of the increasing percentage of carbonic acid [carbon dioxide] in the atmosphere, we may hope to enjoy ages with more equable and better climates, especially as regards the colder regions of the Earth, ages when the Earth will bring forth much more abundant crops than at present, for the benefit of rapidly propagating mankind" (Arrhenius, 1908). Thus, the notion that a strengthening of the natural greenhouse effect, due to human causes, might have deleterious consequences for people is a realization that has dawned on us only slowly. A train of thought has evolved – in scientific circles, in the public mind, and in the policy world – concerning what those consequences might be. That evolution of thought continues.

The greenhouse effect is a complicated problem and a global problem. Almost everybody contributes to the greenhouse effect. Even if the United States somehow could stop producing CO_2, the world would still have a global warming problem. That is not to say that the United States need take no action. On the contrary, the United States is so wealthy and influential that it may be able to do more than any other single country to lead the world toward a sustainable future. American initiative is essential, but global problems require global solutions, and all countries will have to play a part.

14
The Keeling Curve

In the Preface of this book, I have mentioned the Keeling curve, the most famous graph in all of Earth science, which is Figure 14.1. This graph is the result of the persistence, vision, and skill of Charles David Keeling (1928–2005). He spent nearly his entire career at Scripps Institution of Oceanography, now part of the University of California, San Diego.

Charles David Keeling, known to many of us as Dave, was a wonderful scientist whom I knew well for the last 25 years of his life. He had a very original mind. I recall writing a research proposal together with him. A research proposal is a request scientists send to a funding agency, such as the National Science Foundation, describing research they would like to do and asking the agency to approve a grant providing the necessary money to do it. Dave Keeling told me he thought that we should collaborate on writing our joint research proposal by talking on the telephone, although our offices were located close together. Keeling had the idea that if we worked on the proposal together in his office, or in my office, our colleagues passing by in the hall outside the office would think we were just having a conversation, and they would feel free to come in and interrupt us. We therefore spent about a week talking on the phone with one another, writing our research proposal. Indeed, Dave was right. Nobody interrupted us.

Keeling earned a PhD in chemistry at Northwestern University and then held a postdoctoral fellowship at the California Institute of Technology. That is where he began measuring atmospheric carbon dioxide amounts, using an instrument he designed and built himself. He went outdoors to a state park to find unpolluted air. He describes this work in his autobiography (Keeling, 1998):

> Not being sure that the CO_2 even in pristine air next to the Pacific Ocean would be constant, I decided to take air samples every few hours over a full day and night .... Why did I devise such an elaborate sampling strategy when my experiment didn't really require it? The reason was simply that I was having fun. I liked designing and assembling equipment.

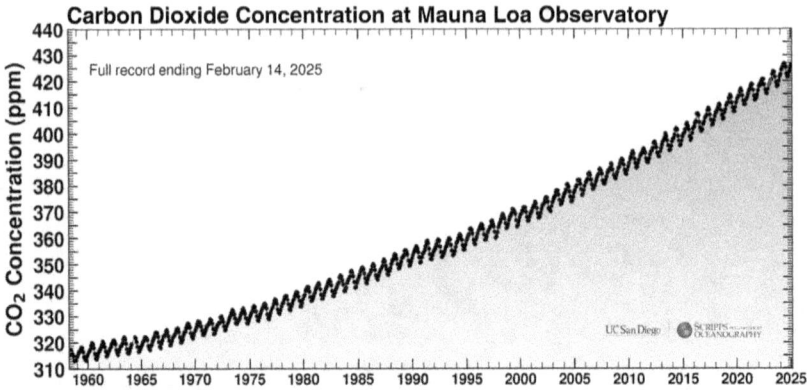

Figure 14.1 The Keeling Curve, showing atmospheric carbon dioxide amount from 1958 to early 2025 at Mauna Loa in Hawaii. These measurements begun in 1958 by Charles David Keeling are still being continued, and this chart is updated daily by Scripps Institution of Oceanography, University of California, San Diego.

Figure 14.1, the wiggly Keeling curve (scientists call every graphed line a curve, no matter what its shape) shows the results of measuring atmospheric carbon dioxide concentrations or amounts since 1958. The curve is now the result of measurements made by many people. But originally, and for many years, the work was carried on entirely by Keeling himself. He realized that it needed to be done, and he created the technology to do it. There was no instrument to measure CO_2 accurately until Keeling invented one. Keeling showed the rest of humanity that CO_2 can be measured accurately, that it is increasing, and that the increase is due to human causes, mainly burning fossil fuels: coal and oil and natural gas. Keeling received many honors and awards for his research, and his fellow scientists regard his work with great respect and admiration. His attention to detail and his passion for accuracy were legendary. His measurements of atmospheric CO_2 are universally acknowledged to be rock-solid. Scientists like to argue – and they do argue about many aspects of the greenhouse effect and climate change – but nobody argues about Keeling's data. Incidentally, there is now another prominent Keeling in atmospheric chemistry. Charles David Keeling's son Ralph also became a faculty member at Scripps Institution of Oceanography. Ralph Keeling is known for measuring the concentration of oxygen in the atmosphere. He also has continued the measurements of CO_2.

The unit for expressing the amount, or concentration, of CO_2 is ppmv, which stands for parts per million by volume, which is equivalent to the number of molecules of CO_2 per million molecules of atmospheric air. The value of CO_2 concentration in early 2025 is approximately 427 ppmv, which reflects a rise of more than 110 ppmv since Keeling first measured it at 315 ppmv in 1958. From analysis of air trapped in ancient ice cores, we know that the CO_2 concentration was about 280 ppmv in the 1800s. So, the increase from 280 to 427 ppmv means that the amount of carbon dioxide in the atmosphere has increased by more than 50% since preindustrial times. We also know from analysis of air trapped in ice cores that it was lower still long ago during periods when much more of the Earth was covered with ice than is the case now. The last glacial peak of the most recent ice age occurred about 20,000 years ago. Thus, we have a very convincing record of the long-term rise of the amount of CO_2 in the atmosphere.

In Keeling's data, incidentally, you can see the temporary slowing of the rise of CO_2 due to an oil crisis in 1973, as well as the effects of natural changes. The recurring wiggles in Keeling's CO_2 curve are evidence of the natural annual cycle of carbon in the biosphere, the world of living things. Plants, mainly the trees on land, account for this annual cycle. Because there is far more land in the northern half of the globe than in the southern half, most of the trees and other plants are found in the northern half of the planet, and thus Keeling's curve follows the Northern Hemisphere seasons. During the Northern Hemisphere spring, plants take in carbon in the form of CO_2 from the atmosphere, thus drawing down the CO_2 concentration and producing the low points on Keeling's curve. In the Northern Hemisphere fall, when these plants shed their leaves, they give off carbon and the curve goes back up again. This rhythmic, seasonal rising and falling of the curve may be thought of as the biosphere breathing, the trees taking in and subsequently giving out carbon in a yearly cycle. It is a clear example of how living things, plants in this case, influence the global environment.

The most significant aspect of the Keeling curve, for its implications for the greenhouse effect, however, is its steady upward climb in recent decades, due primarily to the emission of CO_2 into the atmosphere whenever we burn coal, oil, oil products such as gasoline, or natural gas. Today, fossil-fuel combustion is the main source of the increase in CO_2 in the atmosphere. Deforestation, arising from the cutting down or burning of rainforests and other wooded areas, is a secondary source. In the past, when humans did not burn so much fossil fuel and when there were more forests, deforestation was the primary source. There are significant sources of CO_2 from several other human activities as well, including cement manufacture.

The fuel that we are burning today was formed in the distant geological past from the carbon-rich remains of plants and animals (hence the term "fossil fuel"). It is in this sense that Revelle and Suess (1957) wrote of our doing a one-time geophysical experiment. Once we have taken the fuel that is stored up as coal, oil, and natural gas and burned it, thus putting the carbon back into circulation, we cannot do it again. It will have been used. This fossil fuel took millions of years to form, and we can burn up much of it in a few centuries – the twinkling of an eye on the geological time scale. Thus, fossil fuels are truly nonrenewable resources.

The approximate annual global production of carbon dioxide is an amount we know. We do not have highly accurate estimates of the amount of CO_2 added from deforestation and other land-use changes, but we do know how much coal, oil, and gas is extracted from the ground and ultimately burned. From fossil fuels, we are currently putting about 7 billion tons of carbon in the form of CO_2 into the atmosphere every year – about 1 ton for every person on the Earth. Only some of the carbon produced by the burning of fossil fuels contributes to the increase in the measured atmospheric concentration of carbon dioxide. The rest is absorbed by the ocean and by the terrestrial biosphere, that is, the world of living things on land, such as trees.

Let's look again at the composition of the atmosphere. It is important to keep these facts in perspective. Dry air is a mechanical (not chemical) mixture of gases, of which the most abundant by far is nitrogen (78%). The second most abundant gas is oxygen (21%). Thus, together they make up about 99% of the dry atmosphere. Argon (chemical symbol Ar), an inert gas, accounts for most of the remaining 1%. All the greenhouse gases, except water vapor, are very rare relative to those three gases that dominate the composition of dry air.

Real air is made up of not just dry air, of course, but also water vapor, which is present in natural air in highly variable concentrations. There is hardly any water vapor over the desert, for example, and there is a great deal of it over tropical rainforests. It is also much more abundant near the Equator than near the poles. In the tropics, water vapor locally may be as much as 4% of the atmosphere by volume. Near the poles, it may be much less than 1%. Even in any one place, the abundance of water vapor in the air can vary a great deal from day to day. Its abundance, as measured by relative humidity, is reported as a standard meteorological observation. (Relative humidity is a measure of how saturated the air is, expressed as a percentage. It is the ratio of the amount of water vapor in the air at some spot compared to the maximum amount that the air could contain at the particular atmospheric temperature and pressure at that spot.)

Water vapor is a powerful greenhouse gas, but nitrogen, oxygen, and argon do not contribute at all to the greenhouse effect. The reason for this lies in the fundamental physics of how gas molecules interact with radiation. To absorb infrared radiation significantly under the pressure and temperature conditions of the Earth's atmosphere, a gas molecule has to have at least three atoms. Both nitrogen molecules (N_2) and oxygen molecules (O_2) have just two atoms and therefore do not absorb infrared radiation. A water molecule (H_2O) has the necessary three atoms to produce a greenhouse effect. The water molecule is thus able to absorb radiation, but only at certain wavelengths.

Other substances, too, have the ability to absorb radiation, but at different wavelengths, so that each absorbing substance has its own unique "radiative signature." These include carbon dioxide (CO_2, a molecule containing one carbon and two oxygen atoms) and methane (CH_4, a molecule containing one carbon and four hydrogen atoms). At about 420 parts per million by volume, CO_2 comprises 0.042% of the atmosphere. That's 42 thousandths of 1%. It is a very uncommon or rare gas. And the amounts or concentrations of the other greenhouse gases (apart from water vapor) are even smaller. Intuitively, it may seem unlikely that a gas that makes up so little of the atmosphere should have such a large effect on radiation, but nature, as John Tyndall found out, is sometimes surprising.

As stated earlier, the present-day (in early 2025) CO_2 concentration is about 427 ppmv, so that out of every million molecules of air, 427 of them are CO_2 molecules. Remember, that figure varies from spring to fall as trees take in and give off CO_2. But it was 315 when Keeling started his measurements. These are well-established numbers and are not all controversial.

The number 280, for the nineteenth-century or preindustrial atmospheric carbon dioxide amount, is a little less certain, because it is measured from air bubbles trapped in ice. The process used to arrive at that figure involves drilling into ice sheets in Antarctica and Greenland, and occasionally into glaciers elsewhere, and bringing up an ice core, a slender cylinder of ice, often many meters long. The core is kept frozen before being crushed and analyzed in the laboratory. The air thus released from the ice was trapped when the ice formed from snow, so it is air from the atmosphere that existed at the time when the ice was formed. The ice can be dated, and its CO_2 content can be measured by shining an infrared laser through the air that was in the ice, to determine its ability to absorb infrared radiation.

That process allows us to calculate the CO_2 concentration in the late nineteenth century and in earlier times as well. We are very fortunate that there are places on the Earth where very long ice core records exist. Without ice cores, we would have no way to know the CO_2 concentration in the atmosphere in the distant past. Since the ice can be dated, we even know what the concentration was thousands of years ago, so there is essentially an unbroken record of CO_2 measurements, from the geological past through the preindustrial period and into the twenty-first century. In 1958, the amount of CO_2 in recently formed ice matches the concentration Keeling found when he first started measuring it in the air. Thus, Keeling's curve, the instrumental record, covering a time period only a little more than half a century, is tied to and extended back in time through this ancient record of data measured from air trapped in ice.

It is interesting to consider that the human-caused CO_2 increase of about 140 ppmv – from 280 ppmv in the preindustrial period to more than 420 ppmv in 2025 – means that about one out of every three molecules of CO_2 in the present-day atmosphere is there because humans put it there. By burning such large amounts of fossil fuels, we humans have indeed significantly changed the chemical composition of the global atmosphere.

The rate of increase of CO_2 depends not only on the amount of CO_2 we put into the atmosphere but also on the carbon cycle, which denotes all the processes that transfer CO_2 between the ocean, the land, and the atmosphere. One of the areas of uncertainty, and thus of major active research, is the detailed nature of the carbon cycle. We know that carbon is transferred from fossil fuels into the atmosphere in the form of CO_2. From there, some of it is absorbed by plants in the process of photosynthesis. And plants, when they decompose, return carbon to the soil. This is one of the ways in which carbon is cycled from the Earth to the atmosphere through living things. Carbon is also continually exchanged between the oceans and the atmosphere.

Thus, carbon is constantly cycling through the land–sea–air system, including the biosphere, the world of living things both in the sea and on land. The rates at which this cycling happens, the rates of "flux" or exchange of carbon between oceans, land, and air, are still imperfectly understood. This area of research is complicated by many factors, including the difficulty of measuring the flux of carbon on a global basis. Yet it is vital that we understand more about the carbon cycle if we are to forecast its behavior and its role in climate change.

We have been focusing on carbon dioxide, but it is important to keep in mind that CO_2 is only part of the human-caused strengthening of the greenhouse effect. Additional strengthening is due to methane, nitrous oxide, ozone, chlorofluorocarbons, and a few other greenhouse gases, plus some small particles, called aerosols. The growth rates of some of those gases – the rates of increase of their concentrations – are higher than the rate of increase of CO_2 but may be more variable and more subject to deliberate changes in human behavior. For example, if we are successful in phasing out the manufacture of chlorofluorocarbons, their growth rate in the atmosphere will eventually fall to zero.

When we speak of CO_2 increasing the greenhouse effect, then, we are using it as shorthand for a whole suite of gases, all changing due to human activities. Of these greenhouse gases, CO_2 is the most important, both because it absorbs so much of the Earth's infrared energy and because much of the CO_2 remains in the atmosphere for a long time. Some of it remains for thousands of years. Like every other aspect of this problem, the mix of these gases will change as time goes on. Aerosols such as soot (or black carbon) also contribute importantly to the greenhouse effect.

15

The Temperature Record

With some understanding of how the atmosphere's chemical makeup has changed to date, we can now move on to some central climate-change questions. We shall start with a truly fundamental one: Is the Earth warming, and if so, why?

Scientists have been able to compile a record of how the global average surface temperature of the Earth has changed over the last century or so. This record is an estimate; it is not free of errors and uncertainties. How does one estimate the global average temperature? By carefully piecing together records from many thermometers, including measurements made both on land and at sea, correcting them for known errors, and judiciously interpreting them.

The temperature record shows that all of the warmest years on record are very recent years. To be specific, all the warmest years in the entire global instrumental record, dating back to around 1850, are recent years. Before the late 1800s, there is not enough temperature data from credible thermometers located at enough sites to calculate a meaningful global average temperature. Furthermore, the average warming trend in the last 50 years is about twice as large as the average warming trend over the past 100 years, and the warming in recent decades is apparent on all six inhabited continents. In short, there is no doubt that our planet is warming, but all these facts do not tell us why. We shall take up that topic later.

Measuring the temperature of the Earth is neither easy nor straightforward, though there are fewer difficulties today than there were in the past. I want to mention a few of the difficulties to get you thinking about how to interpret this kind of scientific data. One of the things to consider is how much we should trust the results of a certain kind of science. We shall see that neither observational data, such as temperature measurements, nor theoretical results, such as climate predictions, are perfectly reliable. The models that we will look at later, the computer programs on which our climate projections for the twenty-first century and beyond are based, are also subject to considerable interpretation and uncertainty.

Models are uncertain, because the equations we feed into the computer programs are not perfect representations of the way the actual world works. Science is not yet at the point where we understand the climate system well enough, or have the resources to collect sufficient data, to make the equations as accurate as we would like. Thus, the models give us a prediction, and then we have to make a judgment about how much to trust it.

Observational data, on the other hand, would seem to have a certain reassuring solidity. You make a measurement, and you get a number. Is this result a scientific fact, always acceptable as exactly correct, and always measuring what it is supposed to be measuring?

No. Measurements, just like theories, models, and equations, are subject to uncertainties for lots of different reasons. Perhaps a particular thermometer was inaccurate. There are also many other possible reasons that are far more subtle than that.

The measurements on which we base our estimates of the Earth's global average surface temperature are contaminated by several serious sources of error. One of them is sampling error: We have not made enough measurements, uniformly in space and time, with identical instruments at identical locations on the Earth over all of the 100-plus years of record. About 70% of the Earth is covered with ocean, and the Southern Hemisphere is nearly all ocean. For many years, very few measurements were made at sea because people do not live there. And even within the remaining 30% that is land, there are vast areas, such as the immense ice-covered areas of Antarctica and Greenland, where very few people live and very few measurements were made until quite recently.

The climate record includes measurements not only of air temperature but also of ocean temperature. The early part of the record of sea-surface temperature comes from very sparse measurements made from ships – originally sailing ships, and later motor-driven ships. Those data are also contaminated by changes in the way measurements were made as time went on. The earliest measurements at sea were made simply by hoisting a bucket of sea water up on the deck and putting a thermometer in it. Later, people measured the incoming temperature of the sea water used to cool the ship's engine, using instruments placed under the water line at the cool-water intake. Moreover, ships do not travel everywhere on the ocean. They tend to follow well-established routes. As shipping lanes changed over the years, more geographical sampling issues were introduced into the measurements. More recently, satellite measurements and data from the Argo program, discussed earlier in this book, provide greatly improved ocean data.

Air-temperature measurements are also not immune to error. Measurements on land are made largely from the network that is in place to take daily observational data for weather prediction purposes. Many of those measurements are made at airports, but temperatures at many airports have increased artificially as the cities around the airports have grown. This phenomenon is known as the urban heat island effect, and it arises because city streets and buildings absorb sunlight more effectively than does bare or vegetated land. The airport that was originally built outside of town is often now well within the city's developed periphery, resulting in higher temperature readings.

There are many other ways in which these measurements may have been contaminated – by instrument errors, sampling errors, and plain old mistakes. Some of these occur despite great technological sophistication. In the modern record, people involved in weather forecasting, before they use data, run them through a complicated series of quality checks.

Processing the data used by weather services as a basis for the daily forecast involves passing the measurements through several filters, checking them for internal consistency. As time has passed, these quality checks and filters have evolved and improved.

These comments apply to other climate measurements as well. This means that, like the results from theories and models, measurements need to be questioned. They need to be interpreted. They need to be treated intelligently by people who know how they were created. If you do not understand where the numbers came from, you are more likely to make errors in drawing conclusions from them.

Thus, the task of scientists who study global warming includes analyzing the sources of errors and uncertainties, correcting for the factors that can contaminate the measurements, and using their expertise to develop a best estimate of how the Earth's average temperature has changed, together with quantitative measures of the uncertainty or probable error in the estimate. For example, comparisons of rural and data show that the urban heat island effect is indeed real; cities are, on average, warmer than their surroundings. However, the influence of this effect on the global average temperature trend has been found to be negligibly small, less than 0.006 degrees Celsius (about 0.011 degrees Fahrenheit) per decade on land, and zero over the ocean, where there are no cities.

In general, the atmosphere over the land has warmed more than the atmosphere over the ocean, and largely because of this fact and the uneven distribution of land and ocean, the warming has not been the same in the two hemispheres. If we analyze the data in still smaller regions, we find that the warming has been different from region to region. And it is clear that the warming has not occurred uniformly through time, either. There are different interpretations and explanations of all these regional and transient aspects of temperature change.

One of the most useful things about climate records is that they can give you an idea of how variable climate can be. Sometimes the best way to think about the future climate is to look at how the climate has varied in the past. Even if you do not understand what caused the changes, you at least gain insight into how variable Mother Nature can be. There is much more to be learned from the global average surface temperature record than just the fact that significant warming has occurred over the last century.

16

Climate in the Future

"It is a test of true theories not only to account for but to predict phenomena," wrote William Whewell, an English polymath and scientist, in 1840. After all this analysis of the past and present, what can we say of the future?

As far as carbon dioxide levels are concerned, we cannot know exactly what will happen in the decades to come. The farther ahead we look, the more uncertain we are about how much energy people will be using and what sources it will come from, and therefore the more uncertain we are about how much higher the CO_2 concentrations will go. We can predict pretty well how much of each energy source we will use next year and the year after that. But ten years from now, who knows? Fifty years from now, who could guess? Who can predict whether nuclear fusion will become an energy source that's reliable and cheap in 50 years? I do not think anybody can. Neither can we predict whether political attitudes will change enough in 50 years so that a country such as the United States will get most of its electricity from nuclear power 50 years from now, as France does today. Nobody knows.

And who can answer one of the most important questions when trying to forecast global production of CO_2: What about the large countries in the developing world? Can anyone predict what China and India will do? The further ahead we look, the more difficult it is to predict what the CO_2 effect on the environment will be. This is because it involves forecasting human behavior, something even harder to predict than the weather.

It seems likely, however, that sometime in the current twenty-first century, CO_2 will have doubled its preindustrial concentration of 280 ppmv. This particular benchmark – the doubling of the preindustrial CO_2 amount to 560 ppmv – has a special role in the scientific analysis of this problem. It is easy for a climate modeler, someone who is carrying on research with one of the big computer programs we will look at later, to just change the CO_2 value in the model, making it twice its preindustrial value. Doing so provides a measure of the sensitivity of the climate system to CO_2. Settling on "doubling" also simplifies efforts to educate governments, the press, and the public.

However, there's nothing magical about this doubling. It is just a benchmark. If a very young boy is 2-feet tall now, he will eventually double his height to 4-feet tall. Later he may triple his earlier 2-feet stature. Those are just benchmarks; there is nothing inherently special about being 4-feet tall, or 6-feet tall. Similarly, CO_2, which has already grown considerably, will indeed double and maybe even triple its preindustrial concentration.

Also remember that some of the greenhouse gases that matter most (including methane, nitrous oxide, chlorofluorocarbons, and ozone) are growing too, some at faster rates than CO_2. The rule of thumb is that right now carbon dioxide is only about half the problem, while the other greenhouse gases and substances account for the other half. So, although we will use the doubling of atmospheric CO_2 as our benchmark, keep these other materials in mind as well.

How will the growth of atmospheric concentrations of greenhouse gases affect the climate? Answering this question is not easy. When we try to foresee the future global climate, we are not simply extrapolating past behavior. We are not projecting into the future the changes we have observed in the past. You cannot look at the child who is 4-feet tall today and predict that he will be 6-feet tall later, simply because he used to be 2-feet tall. But you can use your knowledge of the growth of children and of the parents' heights to predict what his height might be. In the same sense, we are not simply extrapolating a past trend into the future when we make a climate prediction. We are using our scientific understanding of how the climate system works.

Keeping that distinction in mind, then, the typical benchmark figure by which climate scientists now predict the climate will warm in response to a doubling of CO_2 is a range rather than a single number. A range often quoted is 1.5° to 4.5° Celsius (which is 2.7° to 8.1° Fahrenheit). Not long ago, a widely quoted consensus number was the midpoint of this range, 3° Celsius (which is 5.4° Fahrenheit). A similar range, also often quoted, is 2° to 5° Celsius (or 3.6° to 9.0° Fahrenheit). There is more than one way to arrive at such a range, both from observations and from models, and the details are important to scientists doing this research. For other people, it is more important to be familiar with the approximate range, because estimates of climate sensitivity will continue to vary as more research is done.

Arrhenius had calculated a value for climate sensitivity that is somewhat greater than modern estimates. He subsequently reduced his estimate from 6°C (or 10.8°F) to a warming of 4°C (or 7.2°F) for a doubling of CO_2 (Arrhenius, 1908). That latter value is within the modern range quoted in the preceding paragraph. This suggests that the physical basis of how carbon dioxide contributes to the strengthened greenhouse effect was well understood more than a century ago, at least by Arrhenius. It also suggests that the relationship between climate and the greenhouse effect is fundamental and is founded on solid science. However, Arrhenius thought that doubling the amount (or concentration) of atmospheric carbon dioxide might take a thousand years. In fact, it is happening much faster than that. Arrhenius did not foresee the explosive twentieth-century growth in human numbers and in fossil fuel use.

Now is a good time to mention one of the consequences of characterizing climate only by global temperature. It is misleading in many ways to talk about climate change only in terms of temperature change, as misleading as it would be to talk about health in terms of body temperature alone. If you say you have a fever of a degree or two, it may be clear that you are not entirely healthy, but the fever alone cannot identify what is bothering you. We do not know whether you have a mild cold or a serious disease. In the same sense, when we talk about a climate change of 3 degrees Celsius, we can judge that it is more serious than a change of 1 degree, but the single number does not reflect the full nature of the climate change. We have not said anything about other aspects of climate that really matter. We have not talked about whether droughts will be more frequent. We have not said whether the strongest hurricanes will be still more intense. We have not talked about whether sea level is going to rise, and by how much, or what the changes in the patterns of rainfall will be, or what the effects on agriculture are likely to be. We have not spoken about the aspects of climate that matter to human beings and other living things, or the aspects of the economy that are sensitive to climate, especially agriculture. It is thus rather artificial and abstract to talk of climate change simply in terms of temperature.

The idea of using carbon dioxide doubling as an index of climate change is a bit misleading as well. In the real world, we are not instantaneously doubling CO_2, making it twice as much tomorrow as it is today, and then waiting for the climate to come into equilibrium with the atmosphere's new chemical composition. Instead, we are gradually adding CO_2, and many other greenhouse gases and particles, over a period of many decades. And at the same time, we are making other changes to the planet. We are changing the land surface by deforestation and by urbanization and by agriculture, for example. Other global factors may be changing as well. Over a century, the Sun can change its intensity, for instance, or volcanoes and people can add aerosols (small particles) to the atmosphere. Climate researchers have only recently realized that aerosols can have globally large effects.

A lot of things are happening at the same time. The experiment we are conducting on the Earth is not an instantaneous one in which we hold everything else constant and change only the greenhouse gases. Instead, it is a transient experiment in which we are modifying the evolution of a planet, one that would have been evolving into something different even without our help. We know from the temperature record of the distant past that climate is always changing. It shifts on time scales of hundreds of thousands of years between ice ages and interglacial periods like the one we are in now. On shorter time scales, we see climate variations that we cannot explain and cannot predict. We think that many such variations, such as the Dust Bowl of the 1930s in the midwestern United States, are due to the natural variability of the climate.

Hence, in addition to all this natural variability, above and beyond what an engineer would call "noise," we are adding something else to the climate. We are stirring another potent ingredient – additional greenhouse gases – into what is already a very complicated soup. What we are doing is very different from, and vastly more complex than, the clear, simple, idealized experiment of instantaneously doubling CO_2. In short, it is important to remember that using the doubling of carbon dioxide as a sole climate sensitivity benchmark is partially misleading, just as measuring climate by globally averaged temperature alone is also partially misleading.

Keep in mind, also, that the difference between one climate and another can be due to a subtle change somewhere. The difference between a desert and a neighboring rainforest can be due to the presence of a mountain range or a difference in the patterns of storms. Climate is subtle, and big variations in climate can arise from small causative effects. Just as a finger on a trigger can set off an explosion, so a very subtle change in a climate input can cause a massive change in a climate output. There are also processes in the climate system that have an amplifying effect. It is evident persistently when we look at the climate consequences of a change in the greenhouse effect, as simulated by the climate models we will examine.

17
Numerical Weather Prediction

The most comprehensive and ambitious climate models that are in use now at research centers throughout the world are the scientific descendants of models originally developed to predict the weather. Climate models are computer models of the climate system, including the atmosphere, the ocean, the land surface, and other components of the climate system. They have proven to be valuable in understanding and predicting many aspects of climate change. The history of these models is a fascinating one. In a sense, the calculations of Svante Arrhenius described in Chapter 13 may be thought of as a first attempt to model the climate in order to determine the temperature changes that would occur as a consequence of increasing the amount of carbon dioxide in the atmosphere. However, as we shall see, the development of modern climate models clearly began with models to predict the weather. After all, the atmosphere, in which familiar weather phenomena (such as storms) can change over a few days or weeks, is the same atmosphere that may experience much larger changes (such as ice ages) over much longer periods. If we had a complete scientific understanding of the atmosphere, and if we could use this understanding to create a computer model that faithfully resembled every aspect of the atmosphere, then perhaps we would be able to extend that model of a virtual atmosphere to include the ocean and the entire climate system. Our goal in doing this might be to predict not only tomorrow's weather but also the climate of the future decades and centuries. Something similar has indeed occurred. In fact, the research done in creating the first scientific model for weather prediction was done only a few years after Arrhenius (1908) published his final estimate of the sensitivity of climate to doubling the amount of carbon dioxide in the atmosphere.

At this point, we may introduce an important pioneer in this field. Lewis Fry Richardson (1881–1953) was a profoundly original English scientist. He did his most famous work on weather prediction around the time of World War

I. Although his achievements in meteorology were certainly influential and important, he was above all a mathematician and physicist.

Richardson became interested in what we now call computational mathematics while the subject was still in its infancy. He was a pioneer in numerical analysis, which for our purposes means using arithmetic intelligently to find approximate solutions to mathematical problems that are too complicated to solve exactly. Many problems involving calculus are of this type. The reason why numerical analysis is practical today is that we now have computers, which, for the purposes of this discussion, are simply machines that can do arithmetic (addition, subtraction, multiplication, and division) and can store numbers very fast without making mistakes. Thus, if you can cleverly convert a complicated mathematical problem, perhaps involving calculus, into a job requiring only arithmetic, then these machines can help you solve that problem.

Richardson, however, worked long before the digital electronic computer was invented. All his arithmetic, like that of Arrhenius, was done the old-fashioned way, with pencil and paper. He was attracted by the idea of using the newly developed mathematical models to forecast the evolution of the atmosphere – in other words, to predict the weather by solving equations. He thought that this was the way to make the imperfect art of producing weather forecasts into an exact science. His goal was to make accurate weather predictions in much the same way that we make accurate forecasts of phenomena such as eclipses.

Today people do not even seem to think it is remarkable that eclipses are predicted with great accuracy far in advance; we take it for granted. We read in the newspaper or watch on television that there will be a solar eclipse next Tuesday at 10:51 a.m., so we go outside at that time and there it is. That is really astounding, if you stop to think about it. Somebody has figured out how eclipses work and has calculated the orbits of the Moon around the Earth as well as the Earth around the Sun. What is even more amazing is that the forecasts of eclipses are always exactly right.

Why not learn to predict the weather like that? After all, there is nothing magical or occult about the atmosphere. The wind ought to change according to physical laws, just as the relative positions of the Earth and the Moon change according to other physical laws.

Thus, Richardson tried to work out a mathematical system for predicting the weather. He started by doing theoretical meteorology, using equations representing the physical laws that describe the evolution of the familiar meteorological properties such as wind, temperature, pressure, and humidity. These equations involve calculus, and somewhat simplified versions of the equations had already been developed by other scientists and were known by Richardson. Richardson then applied numerical analysis techniques to his equations, converting them in such a way that all that was needed to make a forecast was a lot of arithmetic.

Richardson worked alone on this task while holding down a series of jobs in government, teaching, and industry. In this way, he was very much like Einstein, who developed relativity theory in his spare time while earning his living in a Swiss patent office. In 1914, World War I broke out. A devout Quaker and an ardent pacifist, Richardson was torn between his religious convictions and an intense curiosity to see war close up. Richardson found a way to participate in the war as a noncombatant, by driving an ambulance in France. For two years, as he later said, he was "not paid to think." But he did think. Between trips to the front to evacuate the wounded, he did the arithmetic to make the world's first scientific weather forecast. The job took six weeks. Remember, all the arithmetic – many individual calculations – was done by hand. Richardson's "office" was a series of temporary rest camps near the front lines of battles in World War I.

To understand how his forecast worked, imagine a map of the world with a gigantic grid or chessboard pattern superimposed on it. The squares on the chessboard might be a few hundred kilometers or miles on each side. We will regard this chessboard as a way to describe how meteorological quantities, such as temperature, vary from place to place on the map. The value at the center of each square represents the conditions everywhere in that square. This procedure may be appropriate and justifiable, recognizing that in reality we will never have enough measurements to know the temperature everywhere. For example, thermometers might be located only at observing stations near major cities, separated by large distances.

The starting point of a forecast is a depiction of the meteorological conditions at one specific time, say noon today, Greenwich time. More exactly, what is needed is the temperature, wind speed and direction, air pressure, humidity, and so on, at the center of each square on the chessboard. These quantities, determined at the same time for every square, describe the starting point for the forecast. The mathematical term for that starting point is the "initial conditions."

Then the arithmetic comes into play. The equations require knowledge of how much the temperature and other conditions change from one square to the next, from west to east and south to north at all of the squares on the chessboard. These changes, called gradients, are calculated from the initial conditions. From a physical point of view, without doing any mathematics, we can easily appreciate why the gradients are important. For example, one way that the temperature at a given point, say Munich, Germany, can change is if the wind blows colder air into the region. The term for that process is temperature advection. On the chessboard covering a map of the world, as in the real world, if we know that if it is colder to the north of Munich and if the wind is blowing from the north, then we can expect that the temperature in Munich will decrease, because of temperature advection.

Of course, at the same time something else might be happening to make Munich warmer instead of colder, such as the Sun shining down brightly on Munich all day long. To calculate whether the day will be sunny, we need to know whether clouds will form over Munich. To solve that problem, we need to know many other things, like humidity. And just as knowing the temperature gradient is important in predicting temperature advection, we need to know the humidity gradient to calculate humidity advection to predict humidity to predict clouds to predict sunshine to predict temperature changes.

This example is typical, in the sense that all of the quantities that affect weather are connected to one another in complicated ways. You can easily see that it takes a lot of data and a lot of arithmetic to make a single weather forecast using a model like this.

Richardson's first numerical weather forecast, obtained after a great deal of tedious calculating, was not very accurate. In fact, it was wildly inaccurate. The changes that he predicted would occur over six hours at a point near Munich did not occur. In fact, these changes were impossibly large and could never have occurred. Richardson suspected that the error was due to poor observations, leading to inaccurate initial conditions of the forecast, and in hindsight we know that he was right. The observing stations were neither numerous enough nor sufficiently close to one another to provide representative values at the center of all the squares. In addition, the atmosphere is three dimensional, and conditions at higher altitudes are just as important to the forecast as conditions at the Earth's surface. In Richardson's day, there was no accurate way to know what was happening even just above the surface, never mind far aloft.

These observational problems alone would have been sufficient to doom the forecast. In addition, we now know that the mathematical techniques Richardson used were seriously flawed, so that even if the initial conditions had been adequately observed, the methods he used to derive the needed arithmetic calculations would have led to catastrophic errors in the forecast.

There was also one more embarrassing detail. A weather forecast for six hours into the future, even if perfectly accurate, is not particularly useful if it takes several weeks to produce. One feature of a practical weather forecast is that it must be made well in advance of the actual weather.

A less determined scientist than Richardson might have become discouraged and given up. In fact, Richardson did move on to different fields of science, and he has achieved lasting fame for his research in mathematics, in the physics of fluids, and in what today would be called arms control and other aspects of the psychology of peace and war, as well as in meteorology. In all of these fields, his work displays great innovation and creativity.

Fortunately for us, Richardson did one more thing in the area of meteorology. He courageously published his failed forecast. In fact, before he could publish it, he had to find it – it had become lost during the war before finally turning up under a pile of coal. Also, when he did publish his work, he did not confine it to a short article in a scientific journal. Instead, he wrote a book, called *Weather Prediction by Numerical Process*. It appeared in 1922 and is arguably the first modern treatise in the field, now called dynamic meteorology. This book is densely mathematical and not easy going, but it is still well worth reading today. In it, Richardson describes his methods in excruciating detail and reproduces the arithmetic calculations that he had done by hand.

Despite the dismal result of his forecast, Richardson was optimistic about the future prospects of his approach to weather prediction. He knew he was on to something, and he was confident that the observational requirements could eventually be met and that the difficulties that had plagued him could be overcome. He even had a remedy for the problem of producing forecasts faster than the actual weather evolved. Near the end of his book, Richardson describes a vision of a theater-like room holding thousands of people, each doing the arithmetic for one part of the calculation, with a system for transmitting results as needed from one part of the room to another. In modern terminology, this approach to getting arithmetic done quickly is called massively parallel computation, and it underlies the design of some of the most advanced supercomputers of today. In many important ways, Richardson was ahead of his time.

Over the remainder of the twentieth century and into the twenty-first, research developments have unfolded along the lines foreseen by Richardson. Observations improved, as did the theoretical foundations of both meteorology and numerical analysis. The invention of the digital electronic computer supplanted Richardson's imagined theater of people doing arithmetic by hand. In the late 1940s, at the Institute for Advanced Study in Princeton, New Jersey, a team of meteorologists and mathematicians using one of the earliest digital electronic computers produced the first successful numerical weather prediction. By the 1950s, routine weather forecasts were being produced by techniques based closely on Richardson's method.

18
Modeling the Climate System

One of the purposes of this book is to offer insights into how scientists do research. We will turn now to subjects that are a little more technical and a little more uncertain than those we have explored up to now. The aim is to depict science in the process of happening. Current research using climate models is often work at the frontier, the search for knowledge at the boundary between what we think we know and what we do not know yet.

Frontier research illustrates one of the fascinating things about science. Asking questions at the frontier provides some answers but also brings to light new questions. Scientific knowledge is ever-expanding. We never reach the end. We never run out of questions to ask. Also, we never assume that what we think we know could not be overturned by new knowledge.

As we consider how climate models work and how research on the greenhouse effect is carried out, we sometimes find new questions that we now think are important, but that we did not suspect were important only a few years ago. One of the results of hunting for new knowledge is to uncover new areas of ignorance. That is progress.

Albert Einstein was fond of talking about *gedanken* experiments or thought experiments. These are experiments that you cannot actually perform but that are instructive to contemplate. He used them often in thinking about relativity.

The global effect of carbon dioxide (CO_2) in the Earth's atmosphere is often considered by means of a thought experiment. One thinks about the problem and devises a way to find the answer to it, or an approximate answer to it, through a computer simulation. The computer, which was not available to Einstein, serves as a tool to help us carry out our thought experiments. We cannot put planet Earth into a test tube and alter the amount of carbon dioxide on it, but we can create mathematical models of the climate system and use the computer to do such experiments on a virtual Earth.

18 Modeling the Climate System

Earlier, we introduced the pivotal question of what would happen if the amount of atmospheric carbon dioxide were doubled from its preindustrial average. This is a sensitivity experiment. It can help us understand how the climate depends on the amount of CO_2 in the atmosphere. It does not matter how this would happen – we just imagine that we can acquire another 280 parts per million of CO_2 (several hundred billion tons of carbon in the form of CO_2) and dump it into the atmosphere. Then we sit and wait a few decades. We know that the ocean takes a long time to warm up. So, we wait until the climate finishes changing, that is, until it reaches equilibrium with its new chemical composition. Then we look at the new climate and see what the doubling did. That is the sort of experiment we do – on the virtual Earth of a computer simulation, of course. We must always keep in mind that the virtual Earth is not the real Earth.

We will be discussing this thought experiment of climate in a doubled-CO_2 world frequently. Keep in mind that this is different from the climate change that we expect to experience. We are putting the models to an artificial test. The experiment is a benchmark, a way to explore the concepts of equilibrium and doubling. It is not designed to mimic the evolution through time of the real climate system, because in the real world, CO_2 is not instantaneously doubling; it is gradually increasing. It is indeed expected to arrive at double its preindustrial concentration at some time in the twenty-first century, but gradually, not instantaneously. Thus, this thought experiment can give us clues to what the real climate might do.

Moreover, the real climate never comes into equilibrium with the CO_2 that is present, because it is a moving target. That is, while the climate is catching up with one CO_2 concentration, while the ocean is gradually warming, the CO_2 concentration is changing too. The experiment that we are conducting in the real world is a transient experiment, one of continual change, rather than an equilibrium experiment.

Furthermore, we are doing other things to the climate system, in addition to adding CO_2. We are cutting down rainforests. We are cultivating more land. We are paving roads. We are building cities. We are changing the climate system in lots of different ways. Mother Nature is doing unpredictable things as well. The 1991 eruption of Mount Pinatubo, a large volcano in the Philippines, is an example.

If we wanted to design a thought experiment that really did attempt to simulate the evolution of the actual climate, we would need to consider many of these additional factors. Such an experiment might be difficult to interpret. We would be hard-pressed to say which factors were responsible for which changes. Instead, we first try an experiment that may be less realistic, but that may give us more insight into how the climate system works. We simply ask how sensitive the climate system is to doubled CO_2. The difference between such a simple, idealized sensitivity test and a more realistic simulation is an extremely critical distinction to keep in mind. It is typical of the kinds of thought experiments that are conducted not only in climate modeling but also in all kinds of science. For if you try to include everything at once, you cannot figure out which effect is due to which causes.

Subsequently, Richardson's approach was extended to produce computer simulations of climate. Modern research has produced computer climate models in which the atmospheric components of the models are based largely on weather prediction models. Climate models, however, also include simulations of the ocean and land surface and other components of the climate system, as well as the atmosphere. Climate models are also allowed to run for a longer simulated time, several months or even years, instead of the few days needed to produce a weather forecast. This variation in time scales is one of the fundamental differences between weather and climate.

Today's global models for simulating climate change due to the intensified greenhouse effect are the direct intellectual descendants of the hand calculations carried out during World War I by Lewis Fry Richardson. True, Richardson's first numerical weather forecast was ludicrously wrong, and in that narrow sense it was a failure. But he had the right idea, and history has vindicated him. He was a pioneer and a visionary.

An interesting footnote to Richardson's chessboard concept involves the size of the squares, or grid elements. As you might imagine, if the squares are smaller, the simulations of weather and climate can be made more accurate. In the scientific jargon, this is called "increasing the horizontal resolution." But having smaller grid elements, or finer horizontal resolution, requires more arithmetic, so that in practice the affordable horizontal resolution depends on the speed of the computer. One reason why weather forecasts today are more accurate than they were 10 or 20 years ago is that the fastest computers have become much faster. The weather services of wealthy nations have been steady customers for the makers of supercomputers. When a weather service buys a faster computer, it often upgrades its forecasting program to finer horizontal resolution, and the predictions get a little better, just as a more powerful engine improves the performance of a race car. With today's fastest computers, the affordable resolution of global models is considerably better than that of Richardson's chessboard. Typical resolution in 2025 for the best global weather prediction models is less than 10 kilometers, or about 6 miles, as the horizontal size of each grid element. The accuracy and reliability of operational weather forecasts made using the most advanced computer models has improved by about one day per decade in recent years. Thus, a seven-day forecast in the year 2020, for example, is about as accurate as a three-day forecast was four decades earlier, in 1980.

We will introduce two computer terms here. "Numerical weather prediction," the product of Richardson's vision, is the use of a computer model to solve equations by doing a lot of arithmetic to predict the future state of the weather, given the present state. In weather prediction, we start the model at an initial state and run it for a few simulated days or weeks into the future. We are interested in how the details of the model simulation at some future date correspond to reality on that date. That is a measure of the skill of the forecast. "General circulation modeling" employs the same sort of models to simulate climate. This means taking a model, perhaps the very same model as the numerical weather prediction model, and running it to simulate not two or three days or weeks into the future, but two or three months, or two or three years, or two or three decades or centuries. It means simulating the climate by making long runs of the model, thus simulating many days or months or years of weather events, rather than looking at the short-term evolution. These general circulation models, or GCMs, are sometimes referred to as global climate models, and these are the computer programs we are now going to talk about.

There is something fascinating about climate modeling. It is computer simulation of a high order. You can change a number in the computer program and make the Earth spin faster or backwards, you can turn off the Sun, you can make the seasons disappear, or you can change atmospheric carbon dioxide. Of course, you are doing all this with a make-believe Earth, not the real one – and you need to remain aware that it is a mistake to unthinkingly believe everything that emerges from the model is true of the actual Earth. Clearly, there are many things in the real climate system that are not included in climate models. Climate modeling by computer simulation is an exciting area of research that is fun to learn about. It can provide valuable insights into how the climate system works. But the results are approximate solutions of the climate problem, not exact depictions.

As an example, even major thunderstorms are too small to be well resolved by a global model grid. Such storms thus cannot be correctly simulated in these models. They pass through the computed grid the way a small bug can pass through a coarse window screen. So, it is important to keep in mind, when we are discussing global climate, that we are talking about only large-scale features – things big enough to stand out if we looked down on Earth from a viewpoint in space. We are seeing only the big picture. What are small details to a climate modeler are real aspects of climate to human beings. To the climate models, our human ground-level perception of the climate is a sub-grid-scale phenomenon. Much of what interests us about climate is not in these models at all or is addressed only in a highly simplified or idealized way.

Regional and transient effects (in the jargon of climate modelers) are things that happen in limited regions of space and for limited periods of time. They are to be distinguished from the global, equilibrium effects. From the point of view of people, of plants, and of ecosystems, these regional and transient effects are very important. From the climate model's point of view, they are the details. Models that are used to simulate climate change are too broad scaled, too coarse-grained, to represent all the details. In addition, our physical understanding of these regional and transient aspects of climate is imperfect. Research remains to be done. Models are not yet realistic enough to reproduce hurricanes or droughts or monsoons as well as we would like. We should remember that these phenomena, which have major effects on people and ecosystems, may be strongly affected by climate change. We should always be aware that it requires knowledge and wisdom and judgment to understand what model results can tell us about reality.

19
Climate Feedbacks

A little later we will look more deeply into the specifics of how these climate models work. For now, we shall consider some of the challenges the models must contend with. We are going to spend some time talking about feedbacks. If there were no feedbacks, climate modeling would be much simpler, and so would real life.

Feedback is a word that can have several different meanings. In the sense we will use the word in science, it means something very specific. A feedback is a response to a cause, if this response can change what caused it. In other words, a feedback is the modification or control of a process or system by its results or effects. A thermostat can create a feedback if it can respond to a room becoming cold by turning on a furnace that warms the room.

The feedbacks that we are talking about here are ways in which the climate system, as it changes, generates processes that in turn affect the nature of the change. For example, as we change CO_2 – our initial input into the climate system – the climate begins to change, and other things happen as a result. Some of these other things then affect the climate change. These feedbacks are many, and their aggregate effects are complicated. I am going to mention a few of them, and we will talk more about them as we go along.

In a way, completely understanding what will happen when we add CO_2 to the atmosphere is equivalent to perfectly understanding the entire climate system and all of its ever-changing feedbacks. When we are able to say with absolute certainty what the climate will be like at a time late in the twenty-first century, when the atmosphere will contain more CO_2 and other greenhouse gases than it does today, we will have achieved great progress in the science of climate change. We will have an essentially complete understanding of the climate system: atmosphere, oceans, ice, land surfaces, biology, the sun, and all of the interactions that cause climate to change naturally. But until we fully understand that system and its many feedbacks, there will always be uncertainty in predicting what the climate will do. There is also the difficulty of predicting human behavior, if we are to know how much of which greenhouse gases will be emitted in the future. However, we must always keep in mind that a somewhat uncertain answer may still be very valuable. We must not let the perfect be the enemy of the good. Our knowledge of the climate system, like your physician's knowledge of medicine and health, is imperfect but still good enough to be extremely valuable.

The challenge in making climate models realistic is largely that of understanding these feedback processes and incorporating them. As the air warms, does water vapor increase? To what extent? How much does that affect the warming? Do ice and snow melt? How rapidly? Does that change the reflectivity of the Earth's surface and affect the warming? Does the ocean circulation change? In what ways? How does that affect climate? Do the clouds change? Are they more plentiful? Less plentiful? Higher? Lower? Darker? Lighter? How do they feed back on the climate? We need quantitative answers to all these questions. Getting them is the hard part.

If all of these factors stayed constant, if nothing else changed as we changed the CO_2 level, our task would be a lot simpler. But scarcely anything stays constant in the natural world of climate.

Among the most important of these global processes, the feedbacks involving water vapor may seem to be quite straightforward. The underlying concept is simple. Warm air can contain more water vapor than cold air. Air in the tropics should therefore be more humid in absolute terms than air at higher latitudes, and a warmer ocean will evaporate more water into the atmosphere. This is understood theoretically, and it is observed experimentally. Warmer parts of the Earth are indeed more humid; in absolute terms, there is more water vapor in the air. So, as the climate system warms up, we can confidently expect that more water will be evaporated into the atmosphere. Water vapor is itself a powerful greenhouse gas, more powerful than CO_2 in contributing to the Earth's greenhouse effect.

In some ways, CO_2 appears to be simply the trigger for the humidifying of the atmosphere, and models predict that the resulting extra water vapor will strongly amplify the warming caused by the CO_2. Once the CO_2 has warmed the air and the oceans, then the warmer oceans will evaporate more water into the air, since the warmer air has a greater capacity for holding water vapor. The presence of more water vapor in the atmosphere in turn strengthens the greenhouse effect, warming up the oceans even more, and adding still more water vapor to the atmosphere. The result is a positive feedback: an amplifier. Although the term "positive" may sound beneficial, this is not the case at all. Rather, positive feedback connotes a vicious cycle in which a process (in this case, warming) gives rise to effects that strengthen it further. The additional water vapor may also alter clouds, giving rise to still other feedbacks, which we will discuss shortly. The water-vapor feedback itself, however, is treated in models in a somewhat idealized way. Scientists agree that more research is needed on how water vapor changes in different regions of the atmosphere, for example, during climate change. Expert opinion is that both models and observations indicate positive feedback from water vapor, adding to the warming.

A feedback involving ice and snow is the main reason why models indicate that a strengthening of the warming at high latitudes of the Northern Hemisphere will occur, as is observed in reality. Indeed, this feedback is why the warming of the Arctic may be stronger than anywhere else on Earth. As the planet warms, ice and snow melt. The snow line in Canada, Northern Europe, and Asia generally recedes northward toward the Pole. Some sea ice in the Arctic melts, and the area covered by ice and snow shrinks. In place of that ice and snow is now either bare land or bare ocean, which in either case is darker than the ice and snow. Anyone walking to the beach knows that a dark asphalt street is hotter than a white sidewalk on a sunny summer afternoon, because the white surface reflects away some of the radiation from the Sun that a dark surface would absorb. Similarly, where ice and snow might have reflected sunlight away, the darker land or ocean is more effective in absorbing sunlight. So, the poleward retreat of the ice and snow line darkens the surface of the Earth, leading to an increase in the absorption of the sunlight, leading to warmer land and ocean, leading to greater melting of ice and snow. This is another example of a positive feedback, or vicious cycle, where a particular change affects the environment in such a way as to amplify the change itself. We shall say more about this feedback in Chapter 21 when we discuss using climate models to calculate the effect on climate of doubling the amount of CO_2 in the atmosphere.

Recall that characterizing climate change by global average temperature is as misleading as characterizing illness simply by the magnitude of your fever. You might be running a temperature above normal without knowing whether you have a serious illness. In the same sense, whether the planet warms by 2 degrees, or 4 degrees, may be less significant than the consequences of this warming for specific aspects of the planet, such as sea level.

The rising sea level is a serious problem. A consensus scientific estimate is that sea level now (in 2025 as this is written) has increased by about 20 centimeters (or about 8 inches) during the period from 1900 to the present. Sea level continues to rise, and this is a matter of concern for low-lying coastal areas worldwide. Sea level rise will certainly affect vulnerable areas including the Netherlands, Bangladesh, Florida, Venice, many island nations of the Pacific, and the Nile Delta region of Egypt. Because of feedbacks, sea level rise is a complicated problem.

One area of intensive research involves examining sea level rise as a result of feedback processes. Sea level rises mainly for two reasons. First, the ocean expands thermally: Warm water takes up more volume than cold water. That is straightforward – it is an experiment we can perform in the laboratory. Second, ice that is currently on land melts and flows as liquid water into the oceans, also increasing the sea level. In fact, we know from several kinds of proxy data and geological evidence that during the ice ages, when much more water was stored up in glaciers and ice sheets, the sea level was considerably lower than it is today. In fact, during the last glacial maximum, about 20,000 years ago, global average sea level was about 120 meters (or some 400 feet) lower than today.

Alaska and Siberia at that time were connected by a land bridge many miles across. So, you might expect that in a warmer climate, water that is now locked up in ice on land, say on Antarctica or Greenland, will end up in the oceans, and sea level will rise. That is straightforward as well.

But now the story twists and turns. The water story is especially complicated. We are still learning about changes in precipitation and evaporation that will occur as the climate warms. In general, there may be fewer storms in the warmer world, but the stronger storms may generally become stronger. Precipitation will vary. Wet regions may become wetter. Dry regions may become drier. Rainfall amounts may increase. You might say there will be more rain from deluges and less rain from drizzles. We have data, so we should be able to monitor such changes. Figure 19.1 shows the global climatology of

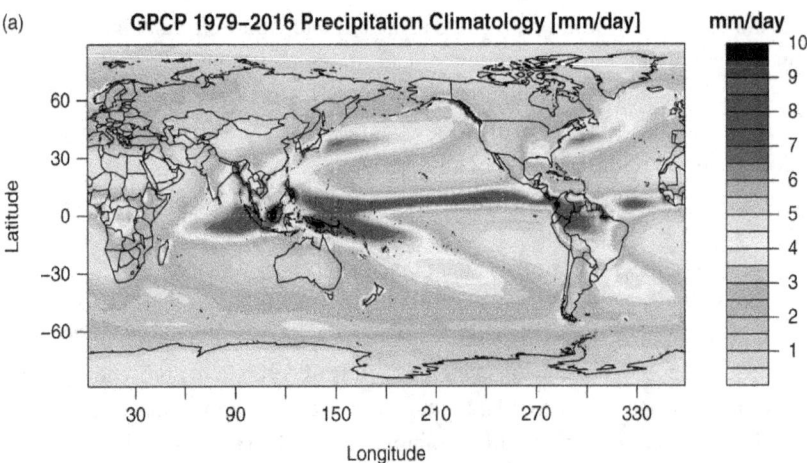

Figure 19.1 Precipitation climatology map (Shen and Somerville, 2019) showing the global distribution of precipitation, based on data from the NASA Global Precipitation Climatology Project averaged over the years from 1979 to 2016.

precipitation averaged over a 37-year period, from 1979 to 2016. Accurate precipitation maps with global coverage such as Figure 19.1 could not be made in the pre-satellite era.

To give an example, one might speculate that the ice sheets of Antarctica and Greenland, while losing mass to the ocean by melting and by processes such as the creation of icebergs, may actually gain mass because the snowfall becomes greater in the warmer climate than before. Thus, in a warming world, one might imagine a situation where sea level rises less dramatically, or even falls, because these ice sheets gain more mass from extra snowfall than they lose from warming. Is this happening now?

Questions like this, involving competing physical mechanisms, must be settled by quantitative research. Observations indicate that changing ice sheets have had a net effect of increasing sea level rise. For example, from 2002 to 2017, according to research based on GRACE satellite data, ice sheets contributed about a third of the total mean sea level rise.

In addition to this hydrologic-cycle feedback and the ice-and-snow feedback, perhaps the most crucial feedback processes involve clouds. Getting a sense of these feedbacks requires an understanding of how clouds affect the planet's thermostat – that is, its ability to regulate its temperature.

Like every other body in the universe that is not at a temperature of absolute zero, both the Earth and the Sun emit electromagnetic radiation. However, they emit at very different wavelengths. Clouds have competing effects, both cooling the Earth by reflecting sunlight (solar radiation) away and also warming the Earth by contributing to the greenhouse effect and thus trapping infrared energy (terrestrial radiation). Thus, both solar radiation (emitted by the Sun) and terrestrial radiation (emitted by the Earth) are involved. The net effect of clouds depends on how clouds react to these two different kinds of radiation. We will return to this subject shortly.

One of the physical laws involved is that, for certain idealized, theoretical entities called "black bodies," of which the Sun and the Earth are good approximations, hotter bodies emit more energy than colder bodies, and they do so at shorter wavelengths. Conversely, colder bodies emit less radiation than hotter bodies, and at longer wavelengths. On the Kelvin or absolute temperature scale, the Sun is about 6,000° on its surface, and the Earth is roughly 300°. To convert from Kelvin degrees (°K or kelvins), to Celsius (°C), subtract 273; the Earth's average surface temperature is about 288°K, or 15°C. Because the Sun has a temperature of several thousand Kelvin degrees, or kelvins, it emits radiation at much shorter wavelengths than that emitted from the Earth.

We use several different names for the Sun's radiation. We call it sunlight, or visible light. Most of it is in the part of the electromagnetic spectrum that our eyes are sensitive to; it is "peaked in the visible," as a physicist would say. Solar radiation is another term for sunlight, and it is also called shortwave radiation. It is what you feel when you go outside in the daytime and the Sun shines on you. You absorb some of that sunlight, and your skin temperature increases.

Most of the radiation that the Earth emits is what we call radiant heat. It is also called infrared radiation, which is just a name for a different part of the spectrum of electromagnetic radiation. You cannot see it, because your eyes are not sensitive to it, although certain instruments are. Within visible light, red has the longest wavelength and blue the shortest. What is longer than red, or infrared, you cannot see, but it is there. It is also called terrestrial radiation, or longwave radiation, because its wavelength is much longer than that of solar radiation. The Earth also emits a little radiation at other wavelengths, notably in the microwave part of the spectrum, but just as the energy emitted from the Sun is peaked in the visible, the energy emitted from the Earth is peaked in the infrared.

In addition, as noted, hotter bodies emit more radiation. Black bodies emit an amount of energy proportional to the fourth power of their absolute temperature, or Kelvin temperature. The fourth power of a number is that number multiplied by itself three times. Thus, 2 to the fourth is 16, which is 2 times 2 times 2 times 2. Similarly, 3 to the fourth is 3 times 3, or 9, times 3, or 27, times 3, or 81. So, if you double something's Kelvin temperature, you increase the amount of energy it emits by a factor of 16, but if you triple its Kelvin temperature, it emits 81 times the energy. (Our understanding of this fourth-power relationship comes from the area of theoretical physics called quantum mechanics. A detailed explanation is beyond the scope of this book.)

Thus, the amount of energy emitted, whether it is heat or light or some other form of energy, depends on the temperature of what is producing it. The amount of energy is approximately proportional to the fourth power of the absolute, or Kelvin, temperature. For the special idealized objects called black bodies, the amount of energy they radiate is exactly proportional to the fourth power of the absolute temperature. As soon as we hear the word "radiation," scientists think in terms of some absolute temperature to the fourth power.

Now it is time to consider the dual role of clouds. First, clouds affect the interaction between solar radiation and the Earth. Some sunlight is reflected by clouds, and part of the sunlight that passes through the atmosphere is reflected by the surface of the Earth. We will address reflectivity in greater depth shortly, but in round numbers, about 20% of the sunlight that hits the Earth's atmosphere is reflected outward by clouds. An additional 10% is reflected by the Earth's surface and by particles in the atmosphere. Thus, about 30% of incoming sunlight is reflected back out into space. When clouds reflect sunlight, they help to cool the planet by contributing to its reflectivity. Obviously, if less sunlight is absorbed by the Earth, the planet is less warm than it would be if more sunlight were absorbed. So, through these effects, clouds cool the Earth.

Second, clouds affect terrestrial radiation. They absorb some of the heat emitted by the Earth, and they re-emit it both upward and downward. Some of the radiation they emit upward is absorbed and re-emitted; some escapes into space. The radiation that clouds emit downward is partly absorbed by the atmosphere and the Earth below. That increases the greenhouse effect. Moreover, molecules of ozone, carbon dioxide, and water vapor can also do just what clouds do. By being partly opaque, by not being transparent to heat, they absorb some of the radiation that would otherwise escape to space, re-emitting part of it downward. That is what we mean by the greenhouse effect. So, through these effects, clouds and greenhouse gases warm the Earth.

In short, clouds affect the Sun's solar radiation in a way that cools the Earth. At the same time, they affect the Earth's infrared radiation in a way that warms the Earth. The same clouds do both of these things.

So, which effect dominates? We can pose that question in terms of another thought experiment. Suppose you could find another planet just like Earth, except that it lacks clouds. A naked, cloud-free Earth. Same sunlight, same oceans. As seen from space, this planet reveals blue ocean, land, ice, snow, and trees, but no clouds. Everything else on that imaginary planet is exactly the same. Will that planet be warmer or cooler than Earth? The absence of clouds would mean that it is less reflective, so more sunlight would get through to the surface. That would help to warm it. But the absence of clouds would also mean a less powerful greenhouse effect. That would cool it off. Which effect dominates?

This question is a nice way to illustrate the importance of numbers to science, to demonstrate that science is inherently a quantitative subject. Lord Kelvin, the distinguished nineteenth-century physicist for whom the Kelvin temperature scale was named, said this about numbers: "When you can measure what you are speaking about, and express it in numbers, you know something about it; but when you cannot measure it, when you cannot express it in numbers, your knowledge is of a meager and unsatisfactory kind: it may be the beginning of knowledge, but you have scarcely, in your thoughts, advanced to the stage of science."

In this case, we are not asking the qualitative question of whether clouds warm us or cool us. We know that they do both. Rather, we are asking the hard, quantitative question: Which of those two events is dominant, the contribution of clouds to the greenhouse effect (warming) or the contribution of clouds to reflecting away sunlight (cooling)? If our actual clouds on the Earth do more cooling than warming, then the imaginary planet without clouds would be warmer. If our actual clouds do more warming than cooling, that cloudless planet would be cooler.

Satellites have been deployed to determine the contribution of clouds both to the Earth's reflectivity and to its greenhouse effect. We can now quantify, from space, those two aspects of clouds. The analysis was done by a team of scientists that included V. Ramanathan, a distinguished scientist who also discovered that chlorofluorocarbons are greenhouse gases.

And the answer to our quantitative question is that, on average, clouds cool the planet. That is to say, the sunlight-reflecting aspect dominates over the greenhouse heat-trapping aspect.

Kelvin says that if we have numbers, we understand something. In the case of the cloud effect, we now have numbers. They're expressed in units of power per area, specifically watts per square meter (W/m^2). These are understandable numbers: incandescent light bulbs typically use 25 to 100 watts of power.

The effect of the clouds on average over the whole planet, in heating the Earth by contributing to the greenhouse effect, adds about 30 watts of power to every square meter of the Earth's surface. The cooling effect of those same clouds, on average, reflects away sunlight equivalent to about 45 watts per square meter. Thus, the clouds' contribution to reducing the amount of sunlight that reaches the surface substantially exceeds their contribution to trapping radiation that the Earth emits. (The figures of 30 and 45 watts are approximate, but they are typical for the global and annual average effect of clouds.)

On our present-day Earth, then, clouds on average cool the climate. So, the imaginary cloud-free planet would be warmer than Earth. However, that planet's temperature variability would also be greater. The days and nights would differ more from each other. At night, clouds obviously do not have an effect on sunlight, but they do have an effect on keeping nighttime temperature higher. For example, the Moon has a far greater day–night temperature difference than does the Earth. The Moon not only has no clouds, but it also has no atmosphere and thus has no greenhouse effect at all.

The Earth has a reflectivity, or albedo, of about 30%. That means that 30% of the energy that strikes the Earth is reflected away. We know this from instruments on satellites. We can also calculate what the Earth's temperature would be if it had that same albedo but behaved like a black body – that is, if it had no greenhouse effect, and its temperature were determined only by receiving sunlight and reflecting 30% of it away. That temperature turns out to be about 255°K. Remember that 273°K is 0°C, that each Kelvin degree, or each kelvin, is the same as a Celsius degree, and that a Celsius degree is a little less than twice a Fahrenheit degree (1.8°F, to be exact). So 255°K is 18°C below a temperature of 0°C and that is approximately 0°F. That is a very low temperature, extremely uncomfortable for humans, and too low for many species of living things.

As noted earlier, the actual temperature of the Earth, averaged over the globe, which we can also measure from satellites and from networks of thermometers, is about 288°K (15°C, or 59°F). That is a much more livable temperature for people. The difference between the actual temperature of our climate (288°K) and what the temperature would be in the absence of the greenhouse effect (255°K) is quite substantial: It is 33°K (33°C, or nearly 60°F). That is a measure of the heat retention caused by the natural greenhouse effect.

We can also calculate how sensitive a black body's average temperature is to albedo. If the albedo were 1% different – that is to say, if it were 31% or 29% instead of 30% – the temperature would change by about 1°K (1°C). So, if the albedo went up to 31%, less sunlight would be absorbed by the black body, and we can compute that the temperature would go down from 255°K to about 254°K. If the albedo dropped to 29%, the temperature would rise to about 256°K. So, we can say that a planet is sensitive to changes in albedo.

Earth's albedo is due mainly to clouds. Year in and year out, cloud cover on Earth is around 60%, although you will see different estimates. That means that, on the average, 60% of the sky is cloudy and 40% is not. It is actually tricky to determine this percentage, because it is not a simple matter to say whether a cloud is present over a given area or not. You can go outside and see whether the sky is crystal-clear blue or completely overcast. But if you are looking down at the Earth from a satellite, which is how we determine these estimates, and you see a slight haze or some thin cover over a region of the Earth, is that a cloud or not? That is a delicate call, and there are different ways of making that determination. Incidentally, earlier estimates of cloud cover were closer to 50%, because newer instruments are more sensitive to the thin, almost transparent clouds that earlier satellites could not see very well. Scientists differ about how to interpret these data on cloud cover, but for now, we will call it 60%.

That 60% cloud cover is responsible for approximately two-thirds of the albedo of the Earth. Most of the other one-third of the albedo is due to sunlight being reflected from ice and snow on the surface of the Earth and from the bright surfaces of desert regions. This means that we have good reason to think that the climate would be sensitive to clouds. Just a small change in the cloud cover could easily result in a change in albedo of 1%. Thus, even a modest change in long-term average cloud cover could produce a substantial change in the albedo and, hence, in the Earth's temperature.

All these numbers raise some interesting questions. Why is the global cloud cover 60%? There's no universal law of nature that says every planet you stumble upon should have 60% cloud cover. Venus has 100%. Mars has roughly zero. And why should Earth's albedo be 30%? Albedo is not the same on every planet either.

As the climate changes, do the clouds change? Are clouds relatively robust and invariant, and somehow immune to climate change? Were clouds different in the last ice age than they are today? Will they be different in the future? We do not yet have firm answers to these questions.

This may in fact be the most important feedback question of all: How will clouds change as the climate changes? As the climate warms up, will clouds become more plentiful, lighter and fluffier, grayer and darker, higher, lower, more prevalent in the tropics? How will changes in the clouds affect their contribution to reflectivity and to the greenhouse effect? How will those changes feed back into the climate? How will these effects vary regionally?

The biggest single physical uncertainty in predicting climate is thought to be our lack of certainty as to how clouds will change and what their feedback on climate will be. That is important to keep in mind as we turn to the subjects of weather and climate predictability.

20
Predictability

Let us first consider the predictability of weather. After we have gained an understanding of that subject, we may be better able to consider the predictability of climate. Predictability deals with questions of what is possible. It is concerned with learning whether something in particular about the future can ever be known. That is different from the issue of whether we have learned how to predict it. For example, we now know how to predict the times when eclipses will occur, and the times when high and low tides will occur. These events are inherently predictable far into the future. But our ability to make skillful predictions of these events is only a recent development. The fact that we now have the ability to forecast the weather skillfully for a few days in advance is proof that weather must be predictable, at least for a few days, even though our ancestors of a century or more ago had not yet developed a method of successfully predicting it.

You might say that predictability of weather is a question about the natural world, not about human ingenuity. There are thus two questions. How far into the future is it possible to predict the weather? How long are current weather forecasts skillful? To the second question, from three to five days up to about a week or ten days is a reasonable answer. Of course, the morning weather forecast is sometimes wrong by the afternoon. But on average, when you see a weather forecast on television, for example, saying that it will be sunny today and warmer tomorrow and cloudy and cooler the day after, the forecast is right more often than not. There is substantial skill up to about five days, and sometimes for a few days beyond that, but there is usually not much skill at a forecast range of a week or ten days or even more.

That is a statement about how successful people are at forecasting the weather. Part of what causes errors in forecasting the weather is that we do not know today's weather perfectly. We are unable to assemble a complete picture of all the storms in the world at any given moment. We do not know about the little baby storm out in the eastern Atlantic today that might become a hurricane two weeks from now. We are partly ignorant of the initial conditions, every day.

Parts of the world are not observed, and the parts that are observed are not observed in minute detail. If you ask what it takes to make a weather forecast for tomorrow or the next day or the day after that, it is two things. One is knowledge of the weather now, the initial conditions. The other is a means for telling the future if you know the present. In practice, this is a system of computer programs for predicting the future, given the imperfect initial conditions representing our knowledge of the present.

We introduce uncertainty in both areas. We have imperfect initial conditions by not knowing today's weather perfectly, and we use an imperfect process by which we look forward in time. If you ask why the forecast is good for only about a week, the answer is a combination of both sources of uncertainty.

Consider an analogy with the stock market. If you want to predict the stock market, you are spared the first problem, because you know exactly what the prices of stocks are today – these are available in near real time on the internet and from many other sources. Unless you are cleverer than anybody else, however, you do not know the rules by which stock markets change from day to day, or year to year. And because the price of stocks in the future is affected by things you cannot possibly know today, like what the president of the European Central Bank is thinking, then that is a source of imperfection in your stock forecast also.

Weather forecasts today are much better than they used to be, though sometimes it may not seem that way, especially to the casual observer. But people whose economic lives depend on weather forecasts – people like farmers and ship captains and air-traffic controllers – can tell you that today's weather forecasts are indeed much better than those of 30 years ago. Gross forecast errors made routinely in the 1950s and 1960s rarely occur today. And the useful range of everyday forecasts is much longer now than it used to be.

We can document that in great detail. In fact, a leading source of weather forecasts now is an organization called the European Centre for Medium-Range Weather Forecasts (ECMWF). This organization is located in England and continental Europe. It is operated by a consortium of governments. For many important aspects of weather forecasts, ECMWF's predictive skill has improved by about one day per decade in recent years, as mentioned earlier. Thus, a seven-day forecast now, for example, is approximately as accurate as a three-day forecast was four decades ago. Thus, progress in improving forecasts during about four decades has resulted in extending the useful forecast range by about four days. This is a remarkable accomplishment that translates directly into large benefits to weather-sensitive sectors of the economy.

Thus, there is no doubt that weather forecasts have gotten better, and they have done so as both sources of forecast error have decreased. We have better instruments, more advanced satellites, and other improvements for measuring today's weather. We also have better computer programs, faster computers, and more technology for improving the predicting technique.

That is a measure of forecast skill or prediction. It is distinct, however, from what I mean by predictability. The quality of today's weather forecast is a question of how well we humans do at making forecasts. Predictability is a question about the natural system itself.

The way to pose the predictability question in these terms is to ask this question: If we had perfect knowledge of the initial state, if we knew the weather everywhere on Earth today down to every cubic meter of atmosphere, and if we had the fastest computer we could imagine, and if we were so smart that we had completely realistic equations in that computer, including perfectly accurate models of every last detail of how every cloud process works, if all the little nuts and bolts were just right and we were using the same equations that Mother Nature uses, then how good would the prediction be and how long could we forecast skillfully? Seven days? Seventy days? Seven hundred days? Forever?

This is a question about the intrinsic predictability of the atmosphere. How well does Mother Nature herself know what the future will be? This subject may seem arcane, theoretical, even philosophical, but not very practical. However, it turns out to be both fascinating and very practical. The question of how predictable the weather is, in principle, is extremely important to you and me and to the taxes we pay.

The answer is that we think that the weather is predictable up to about three weeks. We are not certain that the predictability limit is exactly three weeks – it might be 14 days or 18 days or 25 days or 30 days. It varies too, and weather on some days is more or less predictable than weather on other days. However, we are certain that there is a limit, and that it is not seven days, and it is not seven months. It is a few weeks. That predictability limit comes about for the following reason. The atmosphere is not a regular, cyclical, periodic system, like the Earth orbiting the Sun year after year, following the same path, taking exactly 365 and a fraction days, or like the Earth rotating around its axis in very close to 24 hours.

The atmosphere is somewhat erratic. It has a random component to it. It is unstable, the way a stream flowing down a mountain is unstable. If you put a leaf in the stream at one place, it might end up on one bank. If you put the leaf an inch away, it might end up on the other bank. And if you put the leaf at either place a few minutes later, it might end up somewhere else altogether.

The philosopher Heraclitus said, "You cannot step twice into the same river." Or think of an old-fashioned pinball machine, the kind where you pull the handle, and the ball rolls down a track and hits a peg. Depending on how it hits that peg, it then goes either left or right. A very small change in the initial velocity of the ball makes a big change in where it ends up.

Mathematicians call this phenomenon "sensitive dependence on initial conditions." It is exactly how things happen in the pinball machine. If the velocity of the ball did not depend so sensitively on the initial way you pulled back the spring and let it go, you could make the ball follow the same trajectory time after time. It is the same with a golf shot. A tiny difference in the way you swing the club makes a huge difference in where the ball ends up. Even the most skillful golf players are unable to reproduce exactly the same result, shot after shot. That failing is clear evidence of sensitive dependence on initial conditions.

We think the atmosphere is unstable in that same way. We will never know exactly what the initial state is, because it can be disturbed, to put it rather poetically, by the simple flap of a butterfly's wings, and can end up in a very different state.

The flap of a butterfly's wings is a metaphor used by a scientist named Edward Lorenz. The metaphor states that if there were a planet exactly like the Earth, except that it had one additional butterfly, and that butterfly flapped its wings just once, then eventually the weather pattern on that other planet would look very different from the weather pattern on Earth.

That may seem farfetched, but as far as we know, the theory is valid. What it means in terms of practical forecasting is that any error in specifying the initial conditions, and we know errors are inevitable, is going to make the forecast go wrong after a certain time.

Sensitive dependence on initial conditions means that if the other planet, the one where the extra butterfly flapped its wings, is very close to Earth at the initial time, you might expect its weather to follow closely to the Earth's weather for a while, but eventually the two weather systems would diverge, until at some sufficiently long time in the future, the weather on one of the two planets would not resemble that on the other at all, or at least not any more than two randomly chosen states of the atmosphere of a single planet would resemble each other. That is what we mean by sensitive dependence on initial conditions.

The atmosphere is not periodic. It is irregular. There are an infinite number of possible configurations of the atmosphere. A storm could be anywhere. The temperature of any given parcel of air is variable too, as are the humidity and wind speed and direction.

Just as two planets that were almost identical at some initial time would eventually evolve to having quite different weather systems, every weather forecast for Earth will eventually go bad. Even if the forecast were made with the perfect rules, even if we knew the equations that Mother Nature uses to make the atmosphere evolve, even if we had all the physical laws nailed down and a computer program that solved them exactly right and a computer that could handle all the arithmetic and make no mistakes, the forecast would eventually become inaccurate.

Remember Kelvin's remark, that when you can describe something with numbers, you understand something about it. I think what Kelvin really meant was that for many things we are interested in, it is preferable if we can be quantitative in describing them. If we can say Jane has more money than Judy, we know something. However, if we are sure that Jane's net worth is five million dollars, and Judy's net worth is 500 dollars, we know much more. What is the quantitative answer to how far in advance can we expect the best weather predictions to be skillful? That is to say, can we put a number on this kind of conjecture? How would you go about trying to find that out? I will describe two ways that have been tried.

One is to look back over all the weather maps, all the weather records, which go back many decades. These are the measurements taken to support weather predictions around the globe. It is an enormous data set – the same data used to produce long-term temperature records, plus wind measurements, humidity measurements, pressure measurements, and so on. All these data, taken all over the world for many decades, are carefully archived. If you want to know what the weather map was for yesterday, you can look in today's newspaper. If you want to know what the weather map was for, say, December 23, 1930, you can find it in the archives.

You could take all of those weather maps and search them to find two that look alike. You might even be able to program a computer to find the matching pairs for you. Is there some weather map in the archive that looks just like today's weather map, with the highs and lows in about the same places, in about the same intensities?

If you found a pair like that, one for today, the other for a day way back in the historical records, then you could track the sequence that followed. Suppose that the weather map for September 2, 1941, looks like today's. You could look at September 3, 1941, and September 4, 1941, and see how the atmosphere evolved. You could then compare that to today's weather map and then to tomorrow's and the next day's. By doing this, you would have mimicked Lorenz's metaphor, not by having another planet with an extra butterfly, but by having two states from the archives that looked very much like each other, and then seeing how long, as these two states evolved, they continued to resemble each other.

And in fact, that has been done – there have been very extensive computer searches of the archive of past weather maps. And it has been found that, after just a few weeks, the evolution from one initial state of the atmosphere tends to drift well away from the evolution from a second, but quite similar, state of the atmosphere. After two weeks, three weeks, or four weeks, have passed, even the most similar maps no longer resemble each other. That is one approach to addressing the question.

The second approach is to say, "Okay, let us run a computer model twice." But we do not have a perfect computer model; we do not have a perfect simulation. Weather forecasts usually do go bad after a few days, so we are clearly doing something wrong. Still, let us use the best model we have and hope that it resembles the real atmosphere to a sufficiently close approximation. Let us run this computer model twice, once with one initial state and once with a slightly different initial state.

We can start off with any initial state we want. Remember, the computer program is a simulation of our atmosphere – a virtual weather machine. We then modify the initial state: We can make all the highs and lows slightly higher or slightly lower, or we can speed up or slow down all the winds by a small amount. Basically, we make a new initial state and run it with the same model. This is called an identical-twin experiment, because the same program is used to make the two evolutions. We simulate in the computer, as best we can, the evolution of the weather on one planet and the evolution of the weather on another planet, with the same rules. We then ask how long it takes for those two computer simulations to drift apart so that the two weather maps produced in the two evolutions, after the same amount of simulated time, no longer resemble each other.

The answer is that the two simulations diverge at about the same rate as the weather maps in the archives do. After two or three or four weeks, they no longer resemble each other. The tentative conclusion we draw from this is that the atmosphere has a predictability limit caused by sensitive dependence on its initial conditions, and that limit is a few weeks.

I have said that there are two causes of forecast error. One is not knowing today's weather well (an error in the initial conditions). The other is not having the correct rules to make a prediction (an error in the computer model). It is possible to improve our knowledge of today's weather and thus make the initial state more accurate. You can spend more money to observe the atmosphere better, put up more satellites with more accurate instruments, pay more people to make more measurements using more balloons, buoys, barometers, thermometers, hygrometers, and anemometers. By improving the initial state, you can make the weather forecast better. But there comes a limit beyond which you cannot make it better, because there is a time beyond which Mother Nature herself does not know what she is going to do.

It is not really a butterfly flapping its wings that sets a system in motion, of course. It is the fact that a great big storm today was a little storm yesterday, a tiny storm the day before. The day before that, it was just a ripple that could hardly be seen. The day before that, it was not there at all.

Where that baby storm first occurs is a random event. It depends sensitively on the tiny details of the little ripple. In the same way, the place where the pinball ultimately ends up depends very sensitively on how you start it off. And where the golf ball ends up is determined by exactly how you swing at that precise moment when you hit the ball.

Largely on the basis of this kind of theorizing and these sorts of computer experiments, many countries put together a major effort in the 1970s called the Global Weather Experiment. For a period of about a year, the weather-observing system was significantly enhanced. In addition to the regular observing system, special new balloons, buoys, satellites, and ships were deployed. The weather data system in certain parts of the world was better in 1979, when this effort peaked, than at any time before. When the experiment ended, people stopped making the special measurements. Balloons burst, buoys sank, and the system degraded.

You could clearly see the effects of these enhanced observations in areas of the world where the skill of the daily weather forecast depends critically on knowing what is just upstream. ("Upstream" in this sense means the direction from which new storms and other weather systems are coming: In middle latitudes, "upstream" is often just to the west.)

Australia was one such place. To the west of Australia is a great, wide expanse of ocean that had been poorly observed. Before the Global Weather Experiment, Australian weather forecasts often went bad because just upstream was a storm, or even just a ripple in wind patterns, that had not been observed. Without the better instruments, meteorologists often did not know about these events, so they often made poor forecasts. With the improved instrumentation, the forecasts got better. But after the Global Weather Experiment ended in 1979, the forecasts deteriorated again.

We keep very accurate records of weather prediction. One of the reasons why some meteorologists (like me) go into research using theory, rather than doing something practical and useful, like weather forecasting, is that the people in weather forecasting keep score. Every day you know whether yesterday's forecast was good or bad.

The weather services of the world are very competitive, and they all keep records. We know where the best weather forecasts are made. Even in a given forecast office, we know which person is cleverest at interpreting the computer output and making predictions. It is a highly competitive business, and many sectors of the economy are sensitive to it. The governments of the world do not pay for satellites and supercomputers and all the other technology just to please scientists. They do so because weather forecasts have tremendous economic value. The weather-sensitive sectors of the economy, such as agriculture, depend on accurate weather forecasts. And it is easy to demonstrate that the payoff is real. If we put a dollar into making the weather forecasts better, we get several dollars back in increased economic benefits. That is why governments are willing to make these investments in satellites and supercomputers.

Experience has shown that predictability theory has paid off. Predictability theory not only told us that there is a limit of a few weeks on how long we could ultimately hope to make useful weather forecasts, it also told us something quantitative about how much we could improve if we got rid of some of the causes of error. You could do the experiment, for example, with a certain difference in the initial state, or with another difference in the initial state. If you pay for only a limited number of weather observations, your initial error is bigger. The forecasts decay faster and become useless sooner. So, you can make a trade-off. You can decide how much you are willing to spend to improve the forecast. As you do the experiment, you can refine your estimates of how sensitively the forecast skill depends on the initial error.

It has also turned out to be extremely useful to make more than one forecast, explicitly recognizing that the error in the initial state (or the forecast model) causes errors in the predictions. Thus, by making an ensemble of forecasts that differ slightly in initial states (or model properties), one can improve on the skill of a single prediction.

There is another very practical consequence to having this understanding of the limits of forecast skill. It provides you with a way of spotting quacks. If someone says, "I know what the weather is going to be on inauguration day after the American presidential election next year," you know that it is a fraudulent forecast, because it is impossible in principle to make a detailed forecast of the weather three weeks or more ahead. Just as there are certain equations that have no solution, there are questions to which the answer is, "You will never know." You will never know a year ahead whether there is going to be a thunderstorm over Paris, France, at 3 p.m. on July 14. Such a detail of the weather is inherently unpredictable.

As far as our knowledge goes today, everything I have said about predictability is absolutely true. The remaining questions are largely quantitative. Can we sharpen the rough three-week estimate of predictability and get it more exact? Can we tell what the limit is for predicting thunderstorms, versus the limit for hurricanes, versus the limit for tornadoes, versus the limit for very large, very broad-gauged events that last longer and have a longer predictability time? Does the three-week limit apply only to certain larger-scale features of the atmosphere? Are there exceptions to these rules?

Many interesting questions remain, but the basic concept seems solid. As for our current forecasting skill, we can predict weather for at least several days. For the largest-scale features – very large weather systems, highs and lows on continental scales, the things that dominate patterns of atmospheric pressure – the standard is thought to be around a week. On a weather map, these large-scale patterns look like big fingerprints and are represented by isobars, lines connecting all the places with the same atmospheric pressure. These are the things we would hope to be able to predict accurately for a few weeks. We are not there yet. But at least we have some skill in weather forecasting, our skill has improved in recent decades, we know some ways to improve the system further, and we have some reason to think that extending our forecasts might be possible.

None of this, however, says anything about climate. There is a great difference between the predictability of weather, as we have just discussed, and the predictability of climate, which is what we are trying to do when we forecast the consequences of changing the greenhouse effect. Remember, climate is the sum total of weather over great expanses of time. Climate is average weather but also extremes of weather and probabilities of particular weather features. We have not talked about whether it is possible in principle to predict that next summer is going to be warmer than usual in Western Europe, for example. That is a different kind of concern than the forecast of a thunderstorm this afternoon near Paris. Also, some kinds of long-range predictions are trivial to make but useless, such as saying July in Paris will be warmer than January.

These big questions, like how much warmer the planet as a whole will be 10 years or 100 years from now, may in fact be answerable because they represent a different kind of question. It is a question about the statistics of weather, about the overall effects of weather – about the totality that is climate rather than the specifics of weather.

So, we are left with an unanswered question: How predictable is climate? What features of climate are predictable, and at what time ranges? Although we do have some evidence that weather is predictable, when it comes to climate we do not have very much in the way of demonstrated skill. We have hints of skill at monthly and seasonal ranges, and some forecasts of climate changes such as the global average of the surface temperature of the Earth over several years have turned out to be skillful, but we have no track record at all in making predictions of climate change for, say, a century ahead.

When we make a forecast (or a scenario or an outlook or whatever other name we choose to give it) for the climate of several decades, say for all the remaining years of the twenty-first century, we are doing something we have never done before. It is as if we went to the racetrack and somebody said, "I have picked the winning horse in the next race. Place your bet on this horse." You might say, "Hmm, before I risk my money, I would like to know something about your experience in this field. How well do you do at predicting horse races?" At this point, that person might be honest and say, "Well, I have studied this subject a great deal, I have theorized about it, I have amassed a good deal of data, and I think I have come to a considerable understanding. But this is the first actual horse race I have ever tried to forecast."

That is something to keep in mind. This is a new field. We do not have a long track record in successfully predicting the long-term evolution of our climate. We do not have a theory of the predictability of climate that is anything like as firmly grounded as our theory of the predictability of weather. We do not yet have even a firm conceptual basis for knowing specifically which aspects of climate might be predictable and which might not.

Despite these barriers, however, recent progress in research has led to major advances in our understanding of climate. These advances have greatly increased the confidence of scientists in their ability to make skillful and useful forecasts of how the climate system will respond to increased amounts of greenhouse gases in the atmosphere. We are not deterred by the predictability limit on weather forecasts, because we think that climate prediction is a different kind of problem.

The uncertainty of the forecast of weather derives in part from never knowing the initial state perfectly. Climate prediction, however, might not depend critically on the initial state. That is, the climate of the twenty-first century might depend only weakly, or not at all, on what the climate is at the end of the twentieth century, for example. It might depend more on things like what the surface of the Earth will be like in the twenty-first century. How much of it will be bare ground and how much will be forest? What will the carbon dioxide concentration of the atmosphere be? How brightly will the Sun be shining? These are external conditions, dependent on events that are partially external but that help determine the climate.

In mathematical terms, weather prediction represents an initial-value problem: predicting a future state on the basis of initial values, from what you observe right now. Climate prediction may be more of a boundary condition problem: It depends not on the way things were at some initial time, but rather on what is happening at the boundaries of the system. You can think of a boundary condition in simulating the future of a phenomenon as something that is externally specified over the entire time period of the simulation, instead of something that is calculated as part of the simulation. The Sun may be regarded as one part of the boundary of the climate system, because in simulating the changing climate, the climate has no effect on the Sun. The climate of the later years in the twenty-first century may depend in part on how brightly the Sun is shining, on how much energy it is emitting. Carbon dioxide, too, may be viewed at least partially as a boundary, if we think of it not as a basic component of the climate system, but as an external influence that we exert on the climate. The surface of the Earth is also part of the boundary, in the specific sense that air cannot flow through the surface of the Earth. However, the surface of the Earth is indeed part of the climate system in the sense that properties of both the ocean surface and the land surface can and will change as the climate changes. As climate changes, the surface temperature of the ocean certainly changes. On land, ice and snow cover change, and there are many interactions between soil moisture, temperature, and vegetation. But, if you think of deforestation, for example, as something not part of the climate, but rather as an imposed external influence, then cutting down trees in a forest is making a change in the boundary conditions that can affect climate. Pollution particles in the atmosphere, both man-made and natural, may also have a powerful influence on climate: they can reflect or absorb sunlight and can affect clouds. That is now a critically important area of current research, but one that is beyond the scope of this book.

It is important to keep these aspects of prediction in mind. The question of weather prediction is a question of determining the future of a system from knowing its present state. It is an initial-value problem. The predictability of climate, which may well be more of a boundary-value problem, is in many ways still unsolved. We do not know exactly what aspects of climate are forecastable in principle. The climate system has internal variability in addition to being sensitive to external processes, but there are still many good reasons for thinking that important aspects of climate are predictable in principle.

As an example, consider El Niño. What exactly is El Niño? Every few years, unusually warm water appears at the surface of the eastern tropical Pacific Ocean. Water temperature there can be several degrees warmer than normal, disrupting the fisheries off the west coast of South America. The phenomenon often appears around Christmas (El Niño means "The Child" in Spanish) and can last for many months, even a year or more. Research has shown that El Niño is part of a complex of changes in both the atmosphere and ocean, changes involving an area greater than the entire tropical Pacific region and having far-reaching effects, producing drought in Australia and changing the patterns of rainfall in North America. Because El Niño lasts a long time, affects a large area, and produces effects in both atmosphere and ocean, it is not weather. It is an aspect of climate. Recent research suggests that some aspects of El Niño may well be predictable.

Similarly, a specific hurricane is a weather event. But whether hurricanes in a warmer world may be more rare or more frequent, more intense or less intense, or located in different parts of the world than they are now, involves questions of climate. These questions are of central importance in predicting the climate of the future. One of the problems is not knowing which of these questions are answerable in principle. This is a very active area of research today.

We thus have two things to think about. First, how much can we know about the climate of the future? We already have some evidence that weather is partly random. At sufficiently long ranges, weather is unpredictable. Second, once we understand what aspects of climate are predictable in principle, what data and models and other tools must we have to predict them? We need to be dispassionate, detached, and clinical about these questions, answering not with what we hope we can learn, but with what we are able to say is really knowable.

Thus, we would like to know what is predictable about the climate and what is not. Psychologically, you will find that many people want some things to be predictable. While we humans tend to hope there will be at least some degree of free will, there seems to be something deep-set in the human psyche that abhors a completely random future. People want to know what some specific aspects of the future will be, and that craving for some degree of certainty has kept all kinds of soothsayers, oracles, and astrologers in business for a long time.

And after we have figured out what is predictable about climate, we would then like to learn how to predict it. That brings us to computer climate models.

21
How Climate Models Work

Scientists have developed complex computer models to produce climate simulations. We will consider typical models developed by research groups in various countries. Worldwide, there are dozens of research groups that have access to the scientists and the computing power needed to produce these simulations. I will discuss later what goes into the simulations. Let us look at the results first.

The models produce maps that are depictions of how the world of doubled CO_2 in the computer differs from the present-day world. One way to produce these maps is by running the computer model until the simulated climate reaches equilibrium and doing this for two model runs. The first model run is made with the amount of CO_2 in the atmosphere fixed at one number. It might be the measured current amount, as shown in the most recent version of the Keeling curve. This number is not changed during the simulation.

The second model run is made with exactly the same model. It differs from the first run only by having the CO_2 amount fixed at another constant and unchanging number. This CO_2 amount in the second model run is exactly twice the CO_2 amount used in the first run. Each of these two model runs is allowed to continue for a long enough simulated time to allow the model to react completely to the constant CO_2 amount. Thus, the model in each run will produce a climate that changes as simulated time increases. These climate changes will continue to evolve until the simulated climate reaches an equilibrium state, defined as one in which the model climate no longer changes.

By following this procedure, it is possible to create a map of any simulated climate variable, such as average temperature at the Earth's surface, using the results from both of the two model runs. Subtracting the first set of results from the second will produce a map of the changes in the model climate that are caused by the difference in the two CO_2 amounts. Such a map, in other words, thus depicts the difference in the simulated climates between the climate in the doubled-CO_2 world and the climate in the present-day world.

The procedure outlined here thus simulates the effect of doubling the amount of CO_2 and waiting for the doubling to have its effect on the model climate. The global average of the change in surface temperature due to the doubling of CO_2 is a number that is sometimes called the climate sensitivity. For example, if the model with today's atmospheric CO_2 amount produces a climate with a global average surface temperature of, say, 15 degrees Celsius, and the model with twice that atmospheric CO_2 amount yields 18 degrees Celsius for that number, we would say the climate sensitivity for this model is 3 degrees Celsius (or 5.4 degrees Fahrenheit) for a doubling of CO_2.

What do you think the effect of this doubling would be? You might conjecture that various things could happen to the climate. For example, the CO_2 might first raise the temperature a little bit, but then suddenly cause the Earth to be covered by clouds, making it highly reflective, with a resulting cooling that would plunge the Earth into an ice age. That does not happen in any of the models. In fact, all the models predict the same thing: If we add CO_2 to the model atmosphere, the climate warms almost everywhere.

In answer to our original *gedanken* experiment question of how much warmer the Earth would be in equilibrium if CO_2 were to double, several different climate models indicate that, on global average, the planet will be about 3°C warmer. Some models predict it will be only 2°C warmer, and some predict 5°C warmer. In the opinion of the experts who develop the computer models and carry out these simulations, it is very unlikely that the climate sensitivity could be less than 2°C, though it might conceivably be substantially higher than 5°C. However, models with such large sensitivities generally do not match observational data as well as those with sensitivities less than 5°C. (To convert the numbers in this paragraph from Celsius degrees to Fahrenheit degrees, multiply each of them by 1.8.)

Remember that it is best to think of temperature as a symptom – as just one way to measure climate. We are using temperature here as an index. If you move from Denver, Colorado to Los Angeles, California, you can say that the average temperature is higher at your new home than at your old home and that is one way to measure the nature of the difference in climate between those two cities. But there are other ways too: It is more humid in your new location, it does not rain as much in the summertime in southern California as it did in Colorado, it snows in Denver but not in Los Angeles, and so on. Temperature is just one measure.

In our *gedanken* experiment, carbon dioxide is doubled, and you wait a long time – at least several simulated decades, and maybe a few centuries – for all its effects to take place. You wait until the climate has reached a new equilibrium. In this kind of experiment, we are not interested in the transient, real-world picture. This is a thought experiment, not meant to show what will happen in the year 2050 or 2070, but to obtain a measure of the sensitivity of the climate to CO_2.

The models predict the greatest warming due to a doubling of CO_2 to occur in the far north of our planet. We have mentioned this in discussing feedbacks in Chapter 19. Where there are vast expanses of ice and snow – in Canada, northern Europe, Siberia, the North Pacific, the Arctic Ocean, the North Atlantic, and Greenland – the ice and snow start to melt. By the time you wait for all the effects to occur, a lot of ice has melted. A large fraction of the sea ice melts: Half or perhaps three-fourths of it in some places, nearly all of it in others.

In place of the ice and snow that exist in our present climate, in the doubled-CO_2 climate the ice- and snow-covered surfaces have been replaced by darker ocean, bare land, grass-covered land, or whatever was under the ice and snow in the models. All of these features are darker than the ice and therefore exhibit a lower albedo (lower reflectivity). The sunlight hitting the darker surfaces is absorbed, instead of being reflected by the bright-white snow or ice. Thus, there is a positive feedback that occurs in high northern latitudes called, not surprisingly, the ice-albedo feedback. That feedback is mainly responsible for the much stronger warming produced by doubling CO_2 in the far north than elsewhere on Earth. It is called "polar amplification."

Thus, the broad-brush picture you get from these computer simulations of future climate is that, first, it warms everywhere, and second, it warms most in the Arctic, the northernmost region of the Earth. There is not as much warming in the Antarctic, the southernmost region of the Earth, because there is less land there to be snow-covered, and because most of the ice in Antarctica is far too thick for all of it to melt.

One thing that climate models must do skillfully to simulate the doubled-CO_2 climate accurately is to simulate the present climate realistically. For example, the effect of CO_2 in the polar regions is largely dominated by whether there is ice and snow to melt. If there is, then we get a feedback effect, because when it melts, the darker, newly exposed surface absorbs more sunlight. The warming causes a change, and the change adds to the warming. That is a positive feedback, one that amplifies the initial change. (This is the opposite of a negative feedback, which counteracts a change; the thermostat in your house is an example; the thermostat turns on the heat when the house becomes too cold.)

What we are looking at in the polar regions in climate simulations with doubled-CO_2 levels is positive feedback. The change in the ice due to warming amplifies the warming. It does so by removing ice, thus darkening the surface, reducing the albedo, and increasing the absorption of sunlight.

The main challenge of climate modeling is getting the feedbacks right. At present, we cannot blindly rely on the results of climate models, because we are not sure about the extent or magnitude of the feedbacks. If there were no feedbacks, we could determine the Earth's temperature by means of a fairly simple calculation: If you double CO_2, temperature goes up about 1°C. But some climate models suggest that there may be regions in the far north where surface temperatures could increase by as much as 10 times this amount. That is due to very strong positive feedbacks. Other feedbacks come into play, and we will be encountering them as well.

One of the points of this exercise is to show that it is understandable qualitatively, even at the nuts-and-bolts level, by those of us who are not computer experts or rocket scientists or mathematical whizzes. One can talk about these topics in everyday language. True, scientists talk in technical jargon among themselves and use mathematical concepts extensively. However, they do that because the jargon and the mathematics provide a compact and efficient means of communication. It is possible to say the same things in plain language. I have not used equations in this book, and I am not going to. Your physician uses a lot of medical jargon when talking with other physicians, but that same physician switches to plain language when talking to you.

Let us now consider precipitation briefly. For the warmer climates of the future, caused by increasing carbon dioxide, the models generally predict greater amounts of precipitation, relative to that of the present-day climate, in the higher latitudes. In the subtropics, however, which include the Sahara in Africa, the great desert region in the southwestern United States, and their counterparts in the Southern Hemisphere, the models predict reduced rainfall. These changes are due in part to a predicted poleward displacement of storm tracks, the paths followed by migrating cyclones in middle latitudes.

In looking at the complex system of climate, we have really considered only two variables so far: temperature and precipitation. We have not touched on a number of other important variables. We have not asked about wind speed or cloudiness. We have not asked whether the precipitation is coming in heavy deluges or as more frequent drizzles. We have not thought about whether the extra heat appears as severe heat waves or whether each day is just slightly warmer. We are looking at each of these pictures as one slice of reality, one slice of a much more complex computational climate simulation.

Yet we can now begin to piece together what is happening in the models. By comparing the different simulations, we can get a feel for how the climate system seems to work. We are looking at several different representations of the climate system, embodied in several different models. They differ from one another in important respects. But they are all thorough, well-intentioned attempts by hard-working, talented people. They are expensive attempts, too: This is big science, a concept we will return to. Each of the major models uses up many, many computing hours on a multimillion-dollar computer. Yet we see that marked areas of agreement and disagreement among the different models remain.

Still, we can begin to get a sense of what is more reliable in these simulations and what is more conjectural, even though it can be hard to quantify, and hard to put your finger on the details of what seems to be agreed upon. (We are using "details" in the same sense that a scientist would, to indicate regional differences, for example, not to indicate anything trivial.)

I will summarize the models we have examined thus far by noting that there is considerable agreement on temperature change. The world warms in response to increasing atmospheric carbon dioxide concentrations. Different models with different treatments of important feedback mechanisms give somewhat different magnitudes for the warming. There are geographical patterns to the warming, with more warming over the land than over the ocean. There is a distinct amplification of the warming in the far north, because of the ice-albedo feedback. Concerning precipitation, there are also geographical patterns of change.

The spread between the models, the variation in results that different groups of scientists have come up with, is one measure of uncertainty. When all the models agree, they are either all right or all wrong. When they disagree, at most one of them is right, and maybe none is, but in any case there is likely to be more uncertainty where there is broad disagreement. Regional results are clearly more uncertain than global ones, for example.

I think there is a lesson here about the importance of being able to speak quantitatively about the climate. It is one thing to say it will get warmer, or it will get wetter. It is another to say that it will warm by 2°C or 4°C, that it will be 10% wetter or twice as wet, or that all the ice will melt or a fourth of the ice will melt. The quantitative aspects are important. We should keep Lord Kelvin's advice in mind. A great advantage of climate models versus qualitative reasoning alone is that models produce numbers.

Keep in mind also that so far we have avoided discussing the transient picture, the picture of what will happen from decade to decade as we humans gradually change the composition of the atmosphere by introducing more carbon dioxide and other greenhouse gases. Instead, what we have addressed up to now is the simpler problem of climate sensitivity to an instantaneous doubling of CO_2. We have put aside for the moment the question of how fast trace gases will increase. That problem involves predicting human activity, among other things, plus forecasting how the climate will respond to the gradual change.

Instead, we are still using the thought experiment about what would happen if CO_2 were doubled. We are comparing today's world with a doubled-CO_2 world that has been left alone long enough to respond to the CO_2.

Of all the factors that represent the complexity and severity of climate change, global average temperature change is probably the single most important indicator. In that single figure – not just in the patchy fields we have looked at, and not just in the regional picture, but in the global average temperature change – we have found that the models differ in climate sensitivity. Earlier, we discussed a range of climate sensitivity extending from 2°C to 5°C in response to the doubling of CO_2.

So why does science – objective, pure science – come up with different answers to the same straightforward question? Are the scientists confused? What is wrong with their models? Why do they disagree? That is an issue that perplexes science students and policymakers alike. Where does the uncertainty come from? Why is not science able to pin down the answer? Why do we not get one clear answer when we ask these computers what the climate is going to be?

We have done a lot of experimenting with the models, turning the knobs and exchanging the rules for feedbacks and other factors between the models. Models are computer programs, and it is also probably true that these massive and complicated computer programs will contain coding errors, some of which have not been found and corrected, and these coding errors will be different in different models, and they may affect the differences in model predictions. The models definitely differ in how they represent important feedback processes. Quite a few of these processes matter, but one of the most important reasons why the climate models differ in global sensitivity to CO_2 is that they treat clouds differently. The models for which clouds change dramatically, so as to amplify the warming more, might predict a sensitivity of 5°C, sometimes even more. The models for which clouds change very little, or change in such a way as to resist the warming, might predict 2°C.

And the fact is, we are not yet sure which is more nearly correct. We do not even know if that spread of 2°C to 5°C is the best estimate. It is just the range by which some of the most comprehensive and ambitious models, the ones we think the most highly of today, differ from one another.

The effects of clouds and the effects of oceans are thought to be among the greatest handicaps to creating accurate climate models. We simply do not yet have a best way to represent the complicated processes in clouds and in oceans. Keep in mind that clouds contribute to the greenhouse effect by trapping heat, while at the same time contributing to the Earth's albedo by reflecting sunlight away. In the present climate, as we have seen, the cooling effect dominates, but both effects are strong. In other words, clouds have powerful, competing effects on the heat balance of the Earth. That is an example of why climate change can be sensitive to clouds and why clouds can provide important climate feedbacks.

As we have discussed, there are many sensitive questions regarding clouds to which we do not know the answers. We do not know in any fundamental way why the Earth is about 60% covered with clouds. We do not know why something like 30% of received sunlight is reflected off the planet and does not influence climate at all. We do know that these questions are closely related. We do know that approximately two-thirds of that 30% albedo is due to clouds. It is the clouds that dominate the albedo. The clouds are also crucial to the greenhouse effect.

Models use comparably crude representations of the processes that go into clouds. Real clouds are complicated things. There are little water droplets in clouds. There are ice crystals. Hail can form. Snowflakes can form, melting when it gets warm, freezing when it gets cold. All of that is represented in models by gross oversimplifications, partly because the scale of global models is huge: Grid elements may measure hundreds of kilometers on a side, so that each cloudy area in the model represents the effects of many, many clouds. More important, we do not fully understand how all of a cloud's intricate microphysical processes work. That is why the rules for clouds can differ greatly from one model to the next, and why different models exhibit different feedback effects of clouds on climate change. This is now an area of intensive research, with many problems remaining to be solved.

Given that models contain imperfections and uncertainties, how much trust should we put in their predictions for climate change due to doubling CO_2? What is our justification for believing the computer simulations?

My own attitude is that we should take the predictions of the climate models seriously, but not literally. What do I mean by that? The models, though they differ quantitatively to some extent, all agree that the climate will warm substantially, by several degrees, due to a doubling of CO_2. This is a result to be taken seriously. But do not think that every single number in a climate model is going to be completely correct as a description of the actual Earth. That is one reason why when the Intergovernmental Panel on Climate Change, the IPCC, issues reports summarizing the results of research on climate change, it is very careful to indicate confidence limits on each scientific finding, because some research results are more likely to be accurate and trustworthy than others.

Remember, a climate model is actually just a computer program, a set of instructions to a machine, acting on a mountain of real and synthetic data, out of which emerges a simulation of climate in the form of global maps of such parameters as temperature and precipitation. By analyzing the results of many such simulations in detail, we may improve our understanding of how the many different physical processes interact to determine climate. In conjunction with other kinds of climate research, such as observational research to improve our knowledge of the actual, present-day climate, as well as our knowledge of climate in the geological past, computer simulations can lead to insights into how the climate system works. Computer scenarios of future climate, like studies of past climate variability, offer clues about what might happen. Detailed comparisons of computer projections of climate change with the observations of actual climate change can help us make the models more realistic.

The new technology of artificial intelligence is also now being applied to create new and more advanced models of weather and climate. Artificial intelligence is a modern term and often refers to computer systems that can perform tasks that historically have required thought and that only a human could do, such as learning, making decisions, and solving complicated problems. Recent research on using artificial intelligence to improve weather forecasting and climate simulation has already produced new types of models that, at least in some cases, can produce more realistic results while requiring far less computer power than conventional models. This progress relies on technology now being developed and on research now being performed in several countries, involving universities, governmental agencies, and commercial companies. This field is rapidly developing and is very promising.

However, the goal of increasing our understanding need not mean putting absolute trust in the models. Their promise is great. As science advances and experience accumulates, we can expect the models to become better and better mimics of reality. For the present, one of the most valuable roles of models is to allow us to conduct fascinating *gedanken* experiments to guide and stimulate our thinking. Much has been learned from them already, although we are still far from possessing a complete understanding of the climate system and its variability and predictability. Models are part of the scientists' toolbox. Together with other approaches and techniques for investigating climate change, they have already helped to answer one fundamental question: Is climate change in coming decades, due to the human-caused increase in the natural greenhouse effect, going to be real and significant? The answer to that question is yes.

22

The Sixth Assessment Report of the IPCC

As mentioned at the beginning of the Preface of this book, the Working Group One contribution to the *Sixth Assessment Report* of the Intergovernmental Panel on Climate Change, known to climate scientists as the WGI contribution to IPCC AR6, has now been published (Intergovernmental Panel on Climate Change [IPCC], 2021). We should keep in mind that the IPCC reports are mandated to be relevant to policymakers and informative about climate change, but they must not be prescriptive of climate change policy. The working group reports each contain a *Summary for Policymakers*. Each such Summary is not short. For WGI of IPCC AR6, this Summary contains nearly 19,000 words, so it is about as long as a quarter of this book. However, this *Summary for Policymakers* is probably the best relatively concise source of the recent status of climate change science. In this chapter, I have provided my own brief discussion, using nontechnical language, of some of the most important points of this document. This chapter is thus not an official IPCC product. It includes material that I judge will be especially relevant to the topics covered in this book. The language in this chapter is my own, but in writing this account of some important highlights of the WGI *Summary for Policymakers*, I have tried to be faithful to the intent of the authors of IPCC AR6.

The *Summary for Policymakers* begins with a section on the current state of the climate and a statement that human influence has caused warming of the land, the sea, and the air. The report characterizes that statement by using the word "unequivocal" with its connotations of clear, unambiguous, and not in any doubt. The IPCC first used "unequivocal" prominently in its *Fourth Assessment Report* (AR4) in 2007, for which I was a chapter head in Working Group One, and I recall clearly the debate that ensued about using this word.

The IPCC has developed many detailed procedures that are strictly followed, and one such procedure is that the *Summary for Policymakers* must be approved line-by-line in a plenary meeting of the representatives of the governments of all the countries. The IPCC rule is that unanimous consent of every word is required in this meeting, so that if a single country objects to the use of one particular word in a draft version of the Summary, that word will not be in the final version of the *Summary for Policymakers*. The chapter heads of the IPCC report are present in this meeting of the governments, in order to ensure that the results of science as described in the main report are accurately expressed in the Summary, so I was a participant in the meeting in Paris, France, for the Summary of WGI of AR4. This meeting lasted an entire week, and during several days that week, we worked until late at night. The plenary meeting also features formal parliamentary procedure, as well as simultaneous translation of everything said during the meeting into the six United Nations languages. Thus, much discussion of the Summary of WGI of AR4 was devoted to whether various words in languages other than English were sufficiently close in meaning to "unequivocal." This type of meeting can be lengthy and cumbersome, but the result is that when the IPCC assessment report is finally published, no government can say it does not agree with this or that statement in the *Summary for Policymakers*, because it gave its formal approval of every word in the final version of the document, as did all the governments.

The WGI AR6 *Summary for Policymakers* also states that human activities are the cause of the observed increases in atmospheric amounts of greenhouse gases (such as CO_2). The report notes that each of the last four decades has been warmer than any decade preceding it since 1850. The global average surface temperature in 2001–2020 was almost 1 degree Celsius (or 1.8 degrees Fahrenheit) higher than the 1850–1900 average. Additional statements in the report give information about increases in variables such as the retreat of glaciers, temperatures in the upper ocean, and global average sea level. The report concludes that the rate of warming due to human influences is unprecedented in at least the last 2,000 years.

The *Summary for Policymakers* additionally addresses the issue of weather and climate extreme phenomena. These extremes include, for example, tropical cyclones such as hurricanes. They also include cases of heavy rainfall causing flooding, and of inadequate rainfall, causing droughts. Heatwaves are examples of extremely high temperatures. The report states that all of these kinds of extreme events are being affected by climate change caused by human activities.

We have dealt with the question of climate sensitivity in several places in this book. One frequently used measure of climate sensitivity is the global average change in surface temperature that would occur due to the doubling of the amount of CO_2 in the atmosphere. This quantity is often defined to be the value of this temperature change at equilibrium. Thus, it is the temperature change that is reached after all aspects of the climate system have had enough time to react fully to the increased amount of CO_2 in the atmosphere. Climate sensitivity can be estimated or calculated by a variety of methods, which rely on a variety of observations or modeling results. Additionally, sensitivity is often expressed probabilistically as a range of temperatures rather than as a single value.

However, the WGI AR6 *Summary for Policymakers* states that 3 degrees Celsius (or 5.4 degrees Fahrenheit) is now the best estimate of the equilibrium climate sensitivity. It must be said that some expert climate scientists, writing in 2025, have concluded from their research that the equilibrium climate sensitivity is about 4.5 degrees Celsius (or 8.1 degrees Fahrenheit), which is 50% higher than the IPCC AR6 value of 3 degrees Celsius. They may be right. It is clear that the scientific community has not yet reached consensus on this vitally important issue.

In considering future climate changes, the IPCC report relies heavily on the results of climate models using a possible range of scenarios for the emissions of greenhouse gases into the atmosphere. Numerical experiments with such computer models can also include hypothetical changes in other factors, such as the small particles called atmospheric aerosols. For the entire range of possible emission scenarios considered, the WGI AR6 *Summary for Policymakers* states that the global average surface temperature will continue to increase until at least the middle of the current (twenty-first) century. Recalling that the Paris Agreement of 2015 specified a goal or target of limiting warming to 2 degrees Celsius (or 3.6 degrees Fahrenheit), with an even more ambitious aspirational goal of 1.5 degrees Celsius (or 2.7 degrees Fahrenheit), this *Summary for Policymakers* states that neither of these goals will be met during the twenty-first century unless deep reductions in emissions occur.

The *Summary for Policymakers* of the Working Group One contribution to the *Sixth Assessment Report* of IPCC also states that many of the climate changes caused by greenhouse gas emissions cannot be reversed and will persist for hundreds or thousands of years. It cites global sea level rise and the loss of ice sheets and glaciers as examples of changes that will be irreversible on human time scales.

PART IV

The Future

Introduction to Part IV

In 2008, the presidential election in the United States was a contest between two US senators, John McCain and Barack Obama. Both men were convinced that confronting the challenge of climate change was a politically important issue. During the campaign, this issue was often discussed. Of course, the press covered the discussion. As for me, because the office of President of the United States is so important globally, I became convinced that if the newly elected American President spoke out forcefully and made it clear to the world that meeting the threat of climate change was extremely important, there might indeed be significant progress.

For that reason, I published an article in which I imagined the newly inaugurated President making a speech describing the challenge of climate change and promising action. I published the article in the *Bulletin of the American Meteorological Society*, known to us in the field as *BAMS*. This is a journal to which essentially every American meteorologist or atmospheric scientist subscribes. Obama won the 2008 election, but now in 2025, when I am writing these words, it is clear that, at least for those of us who understand the issue scientifically and who think that governments should act forcefully, the action to confront climate change that we have seen has so far been seriously inadequate. We are still waiting for an American President to deliver a speech resembling the one I had hoped for.

23

If I Were President

A Climate Change Speech

By Richard C. J. Somerville

(If I Were President: A Climate Change Speech, *Bulletin of the American Meteorological Society*, **89**, 1180–1182. Published 2008 by the American Meteorological Society. This article from the *Bulletin of the American Meteorological Society* is reproduced by permission)

My fellow Americans:

In my first 100 days in office, I shall emphasize that addressing the challenge of climate change must be a high priority both nationally and internationally.

I pledge today that the United States will play a leadership role globally and will work cooperatively with other nations toward the goal of protecting the Earth's climate for the benefit of all humankind.

In the days ahead, I shall lay out the many benefits for our nation of confronting climate change forcefully, including strengthening national security, promoting energy independence, and increasing economic prosperity and competitiveness.

I shall also outline the threats posed to the United States and the entire world if we continue to give climate change much less attention than it deserves. Americans must realize that the costs of neglecting climate change are very large in comparison to the costs of acting firmly and promptly.

Global climate change is real and serious and ought not to be a partisan issue. The main obstacle to progress is refusing to face reality and dismissing the problem. To produce effective actions for positive change, the nation needs strong leadership and political courage. Climate change is not a problem that the United States or the world can afford to procrastinate about any longer.

If we fail to act decisively, the result will inevitably be a severely degraded climate later in this century. Different parts of the United States in coming decades will be at risk for rising sea levels, decreased water supplies, altered precipitation patterns, floods, droughts, heat waves, and wildfires. The consequences will affect the domains of public health, economic prosperity, and national security.

Making the needed cuts in greenhouse gas emissions will not be possible by any single approach, and all routes to meet the challenge must be explored. The promising options include improving energy efficiency and energy conservation, greater reliance on renewable energy sources, energetic expansion of nuclear power, and development of carbon capture and storage technology. We shall be pragmatic rather than ideological, and we shall favor whatever approaches work well.

This issue begins with the science of climate change. Rapid progress in research has created a body of sound settled science that must be acknowledged and used to inform wise public policy. We know that our climate is currently changing because of human activities, especially the increase in the amount of carbon dioxide, or CO_2, in the atmosphere. This increase in CO_2 is due entirely to human activities, mainly by burning coal and oil and natural gas, and by deforestation.

As a result, the natural greenhouse effect has already been significantly increased by human-caused increases in CO_2 and other heat-trapping emissions. The main immediate result is a warming of the Earth, but many other changes in our climate result from this interference with the climate system. The definitive scientific summary is found in the most recent report of the Intergovernmental Panel on Climate Change – or IPCC – the organization of climate scientists that shared the 2007 Nobel Peace Prize. Among the many scientific results highlighted in the IPCC report are the following:

- The largest CO_2 growth rate in modern times has occurred in the most recent decade. Any attempt to limit further climate change must be aimed at stabilizing the amount of CO_2 in the atmosphere. However, CO_2, the most important of the man-made greenhouse gases, continues to increase in the atmosphere, and more rapidly in recent years than previously. Thus, the trend is in the wrong direction. It must be reversed.
- Our planet is now about 1.4 degrees Fahrenheit warmer than in the late nineteenth century. Observations of many kinds, such as air and ocean temperatures, melting ice and snow, and rising sea level, show that the warming of the climate system is unequivocal. Furthermore, most of the observed warming since the mid twentieth century has at least a 90% chance of having been due to the human-caused observed increase in greenhouse gas amounts.
- The overall warming trend in the last 50 years is nearly twice that for the last century. Tropical cyclones have intensified globally since 1980. Arctic

temperatures have increased at about twice the global rate. Arctic sea ice has shrunk by about 2.7% per decade. The global ocean is warming to depths of at least about 2 miles. The ocean has absorbed more than 80% of the heat added to the climate system. Sea level rise averaged globally over the twentieth century has been about 7 inches. This rate has recently increased.

These are the facts, as established by science. They cannot be ignored.

The main goal in stabilizing the climate must be to quickly reduce the rate at which the global population emits CO_2 and other greenhouse gases into the atmosphere. The speed and size of this reduction will determine how much the climate will warm. Tokenism and good intentions are insufficient. Numbers are what matter: Rapid and quantitatively large reductions in global emissions are essential.

How much warming is tolerable? This is a judgment call, and reasonable people with different tolerances for risk may differ somewhat, but science can help inform the decision. Several other countries have already made decisions about setting a safe limit to climate change. In 2007, the European Union adopted a goal to restrict global warming to no more than 2 degrees Celsius (or 3.6 degrees Fahrenheit) above the average preindustrial temperature level of the mid nineteenth century. In order to fulfil this goal, the European Commission, the executive branch of the European Union, has agreed that developed countries will have to strive to reduce their emissions by 15–30% by 2020. The European Parliament has proposed a European Union CO_2 reduction target of 30% for 2020 and 60–80% for 2050.

Many expert climate scientists would support that decision, and as a first concrete step for the United States, I shall ask the Congress to join me in seeking bipartisan nationwide support for the following 10 urgent actions:

1. Establish concrete goals, timetables, and plans for reducing US greenhouse gas emissions, consistent with those already adopted by the European Union as appropriate for developed countries.
2. Mount a nationwide climate change education program and make clear to all Americans that these goals are the minimum necessary cuts in emissions of heat-trapping pollutants, that even deeper cuts are both desirable and achievable, and that as the level of man-made greenhouse gases in the atmosphere is reduced, the risk of dangerous climate change is correspondingly reduced.
3. Reform the patchwork of federal subsidies, taxes, and other incentives and disincentives so as to encourage large reductions in US greenhouse gas emissions.

4. Emphasize that the first major steps in reducing our nation's greenhouse gas emissions should come from dramatic improvements in energy efficiency and energy conservation, and that these gains typically involve either low costs or actual significant savings.
5. Ensure that our national climate policy is based on sound science and also embraces principles of equity, fairness, and justice, protecting lower-income Americans and those employed in economic sectors adversely affected by climate-change policy.
6. Engage other nations vigorously in a diplomatic effort to find agreement on how all nations can act cooperatively to decrease heat-trapping emissions, in a manner consistent with their greenhouse gas emissions and their state of development and technological capacity.
7. Halt the construction of conventional coal-fired power plants. Instead, encourage the development of large-scale fossil fuel power plants with state-of-the-art carbon sequestration technology, as well as the exploration of innovative large-scale carbon sequestration options.
8. Increase on a sustainable basis the nation's scientific research effort to add to our knowledge of climate change science and the effects of climate change on all regions of the country and all sectors of the economy, while protecting the ability of all scientists to publish and discuss the results of their research freely and without censorship.
9. Revamp our nation's energy policy to immediately reduce and ultimately end our country's overwhelming dependence on fossil fuels, with all its adverse consequences, including the negative impacts of imported oil on the economy and on national security.
10. Partner creatively with the private sector and with states and municipal governments to achieve our national goals for reducing emissions of greenhouse gases.

Additional steps will be needed, but these 10 actions make a good beginning. Climate change presents both a challenge and an opportunity. The solutions must be global, because all nations share the climate, which is a global commons. While every nation will have a part to play, the United States can and must lead in using its immense resourcefulness and creativity to solve the problem.

The time to act is now. Some climate damage has already occurred, and much more is certain to occur, because of the world's collective failure to take adequate actions sooner. Delaying further only increases our costs and limits our options. The choice of what kind of planet Earth we leave to our children and grandchildren is in our hands today.

From this day forward, let no one doubt that the United States recognizes the importance of the climate-change issue and the overwhelming scientific basis for concern. This country will join with others, will lead, and will act vigorously to prevent dangerous human-caused interference with the climate system. The need is urgent. A great nation can meet a great challenge. We start today.

Some Final Thoughts

Looking back at this seemingly naive 2008 article today in 2025, I am reminded that in New York City there is an ancient custom called "Alternate-Side Parking." The meaning of this custom is that parking is not allowed during certain hours on the one side of a specific street on Tuesday, Thursday, and Saturday, and on the other side of that street on Monday, Wednesday, and Friday. That is done for street cleaning and, in winter, for snow removal. Everybody moves their car from one side of the street, often to double park on the other side, depending on what time and what day it is. I have an "Alternating-Side Attitude" about climate change. I am an optimist on Tuesday, Thursday, and Saturday; and I am in deep gloom on Monday, Wednesday, and Friday.

I think, and polls show, that in the United States and many other countries, more and more people are becoming serious about climate change. They understand that climate change is happening. They also understand that human beings have some role in it, but most people usually do not appreciate how dominant that role is. People often do not yet grasp the need for decisive action to mitigate climate change, and many people do not have climate change high on the priority list of things they care deeply about. For these people, climate change may rank much lower than concerns such as national security and the economy and health and education.

As a result, many US politicians do not yet feel any need to emphasize climate change. As we know, some US politicians, especially in leadership roles in the Republican party today, simply do not accept what mainstream climate scientists would consider fundamental, well-supported facts about climate change: that it is real, that it is happening here and now, that it is overwhelmingly human-caused, that it is global in its extent, and that left unchecked, it will eventually have extremely serious, adverse consequences for the world, the beginnings of which we already see. Also, polls show clearly that US citizens who are very concerned about climate change tend to favor or identify with the Democratic party much more than the Republican party. Climate change has become a partisan issue.

Much of the opposition to nuclear power appears to originate from ideological and emotional attitudes, in my opinion, and there is no lack of misinformation and disinformation on this subject. I think that in general, not enough attention has been paid to learning from the experience of other countries that have in several cases tried different approaches to meeting the challenge of climate change.

I think all that is changing, but it is not changing fast enough. I am not sure what it would take to change the attitudes of people who still do not accept the findings of mainstream climate science. At times, I wish for a charismatic figure who could change public opinion the way Nelson Mandela did in South Africa, or Martin Luther King, Jr. did in the US, or Mahatma Gandhi did in India. An inspiring leader can galvanize public opinion. Of course, public opinion can always change. For example, gay marriage is not unusual in the US now, but for most Americans it was absolutely unthinkable until recently.

Climate scientists understand the urgency of mitigating climate change. Climate change is not something the world can safely continue to procrastinate about. We cannot wait until coastal cities become abandoned before we start mitigating sea-level rise. Waiting too long means doing too little and acting too late. There is a timescale built into the climate change issue by physics and chemistry, and the broad public has not yet fully realized the urgency of it.

Reducing emissions of carbon dioxide and other heat-trapping substances will limit the warming. Figure 23.1, which was first produced in 2010, schematically shows three possible trajectories of warming reductions. The area under the three curves is the same, and this area is an approximate estimate of the cumulative amount of carbon emitted that would give us a two out of three chance of limiting warming to 2 degrees Celsius (or 3.6 degrees Fahrenheit) above the preindustrial temperatures of the early 1800s. One curve, labeled 2011, shows that if emissions had peaked and began to decline in 2011, then emissions reductions could be gradual, and by 2050 emissions would not yet need to have entirely stopped. Another curve, labeled 2015, shows that if the date when emissions peak and when emissions reductions begin did not occur until 2015, reductions would need to be much greater, and emissions would have to decrease to zero before 2050. Yet another curve, labeled 2020, shows that if reductions did not begin until 2020, drastic reductions, as much as 9% per year, would have to occur quickly and emissions would need to reach zero by 2040. This graph illustrates the urgency of acting. Similar graphs have been produced by subsequent research based on updated assumptions. Limiting warming to 2 degrees Celsius is the warming target set by the Paris Agreement signed in 2015. At the time of writing these words (2025), emissions have still not yet peaked and begun to decline. They are still increasing. As time goes on,

23 If I Were President: A Climate Change Speech

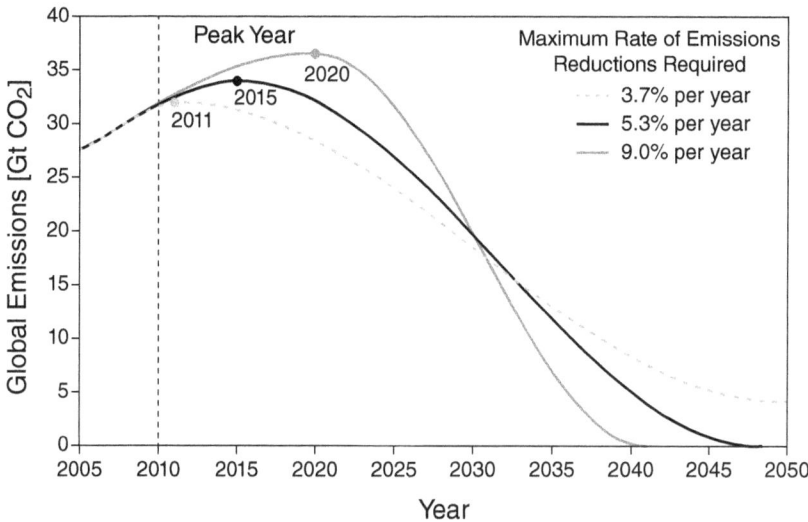

Figure 23.1 Three hypothetical emissions scenarios, each restricting the total global emission of carbon dioxide to 750 gigatons (Gt) over the period 2010–50. If emissions remain negligible after 2050, that restriction yields an approximate 67% probability of limiting warming to 2°C above preindustrial temperatures. The later the year of peak emission, the higher the necessary maximum rate of reducing emissions. The graph was created in 2010, and in 2025 emissions have still not yet begun to decline (Reproduced from Somerville, Richard C. J., and Susan Joy Hassol. Communicating the science of climate change. *Physics Today*, 64: 10, 48–53, 2011, with the permission of the American Institute of Physics).

if emissions reductions still have not begun, it becomes less and less likely that the Paris target of 2 degrees Celsius can be met. The longer you wait before acting, the more drastic the action has to be. The world has procrastinated and dithered for decades, and this chart shows that the result of failing to act is to increase the likelihood of dangerous climate change.

The year 2024 is now the most recent full year of recorded meteorological observations. Recent climate science research on these data has established that 2024 had the highest global average temperature at the surface of the Earth of any year since adequate instrumental temperature data began in the 1800s. In fact, each and every one of the 10 warmest years in this long instrumental record is in the last 10 years. That is to say, between 2015 and 2024, each year is 1 of the 10 warmest years in the entire instrumental record. That is a result announced in a report released in early 2025 (World Meteorological Organization, 2025). It is based on detailed analysis of global mean temperatures in six independent and well-regarded data sets compiled by six different scientific research organizations. The World Meteorological Organization

(WMO) is a United Nations agency headquartered in Geneva, Switzerland. It is responsible for organizing and supporting international cooperation in atmospheric science, climatology, and related fields. Reports such as this one, issued by the WMO, are considered trustworthy by mainstream climate scientists.

The climate problem is inherently urgent. If we do not reduce global emissions of heat-trapping substances such as carbon dioxide – quickly and drastically – then we cannot succeed in obtaining the desired result of limiting climate change to acceptable levels by simply reducing emissions later. It is somewhat like smoking. Waiting to quit smoking, until a lifetime of smoking has left someone in the final stages of deadly lung cancer, means waiting much too late. Delaying too long before reducing emissions of heat-trapping substances will mean subjecting the world to much more severe climate disruption. Mother Nature severely punishes procrastination. In the US, many politicians have clearly not yet understood the implications of that fact.

For mitigating climate change, there is no silver bullet, but there is lots of silver buckshot. There is much that needs to be done. I think scientists do have a role to play in helping to inform the public, but I am not optimistic that progress is happening fast enough. For example, the nations of the world did not agree on how much climate change should be considered to be tolerable and safe, rather than intolerable and dangerous, until the Paris Agreement of 2015 finally settled on an internationally accepted target. The target of the Paris Agreement is to keep global average warming to "well below" 2 degrees Celsius (3.6 degrees Fahrenheit) relative to the preindustrial temperatures of the early 1800s. The Paris Agreement also adopted an aspirational goal to pursue efforts to limit the warming to no more than 1.5 degrees Celsius above preindustrial levels, which is even more ambitious.

I was in Paris in late 2015 when the Paris Agreement was reached at the 21st COP meeting. I was there to assist a group of graduate students from Scripps Institution of Oceanography. I was also accredited to that meeting as a journalist reporting for the *Bulletin of the Atomic Scientists*. The nature of the Paris Agreement is that nations made voluntary pledges of how they would reduce their emissions of heat-trapping substances, by how much, and by when. There is no penalty for failing to keep the promises that the various countries made.

I think the Paris Agreement is promising and is a major accomplishment. It is a step in the right direction, but it does not solve the problem. It provides a mechanism for nations to continue to confer and evaluate the progress that has been made, to monitor the technology that evolves, and to reduce their emissions more strongly in the future. To limit climate change meaningfully, emissions absolutely must be reduced more and more as time goes on. In fact, global emissions must essentially decrease to zero and must do so well within the present century. The world is not on a trajectory now that will lead to a stable climate with only moderate levels of climate change. I hope the Paris Agreement and subsequent efforts will be successful in changing that.

Technologically, I am encouraged by developments such as the rapid increase in using renewable energy sources, especially wind and solar, and the decrease in costs of renewables. I am also encouraged by other developments, such as improvements in nuclear power technology and increases in the rate at which electrical cars are being developed. These are all positive signs. Some countries have made great progress and now develop much of their electrical energy by renewables. This is progress that was unthinkable only a few years ago.

All that is encouraging. Nonetheless, we should keep in mind the warning from Vaclav Smil summarized in Chapter 8 of this book and described in detail in his book (Smil, 2022). He points out that the world relies on producing massive quantities of ammonia, steel, concrete and plastics, "the four pillars of modern civilization." These four pillars all now require very large amounts of fossil fuel energy to produce, and it therefore may require several decades before the world will be able to eliminate its dependence on fossil fuels.

However, we see no great public pressure for more rapid and vigorous action to meet the challenge of climate change. Especially in the United States, we have lacked the needed political leadership. The progress that has occurred in the US has been mainly at state and local levels and in the corporate sector. It has not originated in Washington, DC, the capitol of the country. A great deal remains to be done. I hope the missing charismatic leader arrives very soon.

I also hope the media and the entertainment industry, and particularly the segments that the public pays the most attention to, such as movies and television, embrace this effort much more fully. For example, I would like to see superb climate-change films being made by the real experts, the people who are masters of the art of filmmaking, rather than by amateurs.

In 2025, when I am writing these words, the United States is the only major country in which a current or recent head of government has reacted to the threat of climate change by denying that the problem exists or by characterizing it as a hoax. I think that when a country with the power and visibility of the United States refuses to take a leading role in what many other nations want to do about climate change, it greatly diminishes the chance of success.

Climate change is fundamentally a moral and ethical problem in my view. What do we who are alive today owe to the next generation, and to subsequent generations? I think, at a minimum, we owe them a world that is no worse off than the one we inherited from previous generations. We also owe much to the natural world which today faces many threats. What do we in rich countries owe to those people who do not yet have what we in developed countries would consider the basic necessities of life? These necessities would include access to clean water, health care, education, a degree of security, and not least, some of the prosperity that we in the developed world have achieved, largely by having cheap and abundant fossil fuel energy. At the same time when we realize that today's global fossil fuel energy system must be altered or replaced by a different one in order to reduce the dangers of climate change, we should also realize that the energy system of the future must provide energy that is affordable and abundant. A worthwhile future energy system must provide energy that is not unreliable or intermittent or too costly to build and maintain.

No parents want to see their children inherit a damaged world. To avoid that sad fate, we will not only have to imagine a far less dangerous and far more advanced energy system. We will also have to convince politicians and governments to make the needed changes happen. At present, we clearly have not motivated our politicians and the governments they lead to listen and act, which will require persuading many people to give high priority to the climate change issue. During election campaigns, most people do not yet tell political candidates: "If you want my vote, take climate change much more seriously." I hope this needed change in priority happens. To make it happen may take someone with the moral stature of a Dalai Lama, a Gandhi, a King, or a Mandela. Unfortunately, it may also take something truly frightening that happens, perhaps a sudden destabilization of part of the West Antarctic Ice Sheet. Another possibility is the Atlantic Meridional Overturning Circulation reaching a tipping point that leads to an oceanic instability. Both examples have the potential of major impacts on climate. Many other examples could be cited.

I certainly do not wish for a disaster, but I do wish that the world would wake up. I hope we can awaken the world by logic and persuasion, rather than by watching a real disaster occur. We have already learned a great deal, and the overwhelming scientific consensus is strongly supported by many kinds of evidence. Future scientists will fill in the details, but the big picture has been clear for quite some time.

The 60-year period from 1960 to 2020 has been a time of explosive growth in my field of atmospheric and climate science. This growth has created employment opportunities and research resources for scientists of my generation. This period of rapid growth has ended. Judging by budgets, my scientific field may be in steady state now, or it may well be shrinking. Its future is unclear. Yet we still attract many superb graduate students who are well-prepared and highly motivated. They come to top-ranking academic centers such as Scripps and UCSD because they have a calling to do science. Today, however, I think fewer and fewer of them have their heart set on a career resembling their adviser's career. That is, they are willing to consider careers

23 If I Were President: A Climate Change Speech 159

other than becoming a professor in a research university. Indeed, they must do that, because my academic field has stopped growing.

Many people with PhD degrees in my field will find nontraditional careers. I already know of numerous scientists with PhD degrees from Scripps and UCSD who have become teachers or have joined corporations or have gone into government, and who are doing things other than research. I applaud them. Many of the talented and motivated young people whom we are educating today at places such as Scripps Institution of Oceanography are open to nontraditional careers. Some of these highly trained people with scientific credentials will bring their knowledge and their intellectual ability to careers where we have not traditionally seen many people with scientific PhD degrees from first-rate universities.

I am optimistic by nature. I think if the broad public becomes less gullible and more knowledgeable about science, and if people can more easily spot a quack and tell the difference between real science and fake science, then better-informed voters will help bring about better policy decisions on scientific issues such as climate change. There is a quantitative aspect to all science, and certainly to climate change science. Numbers matter, and some things are just trivially small. For example, some carbon dioxide does enter the atmosphere naturally from volcanoes, but the amount from that natural source is much smaller than the amount due to human activities such as burning fossil fuels. Facts matter too. For example, in the global warming issue, it is obvious that the problem is inherently international. We cannot change the California climate by changing only the California emissions of carbon dioxide. Once we put carbon dioxide in the atmosphere, where some of it will remain for centuries, the winds mix it throughout the global atmosphere. Climate change is an international problem requiring an international solution.

Like all active areas of science, our understanding of the climate system is imperfect. Figure 23.2 illustrates one example of the results of scientific research designed to learn what the future temperature change of the region of the United States will be. This particular example was done using one specific set of computer models of the climate system. It dealt with one specific set of assumptions concerning what the future scenario of emissions that are important in determining the greenhouse effect might be. Different models with different climate sensitivities to emissions and different assumptions about future emissions scenarios would give somewhat different results. All the scientific experiments, however, would predict that larger emissions would lead to greater warming. As our scientific understanding of climate change increases, the quantitative aspects of such experiments will lead to more and more accurate forecasts of the effects of climate change. We learn a little more every day,

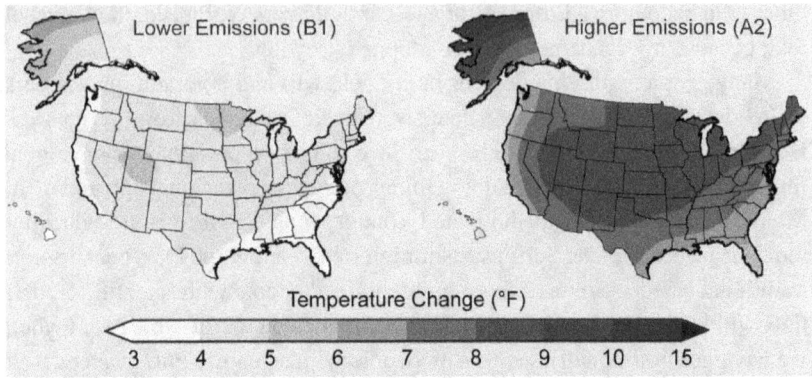

Figure 23.2 Temperature forecasts by a climate model showing temperature changes in the United States resulting from two hypothetical global emissions scenarios. The changes are for average surface air temperatures in the years 2071–99, relative to the years 1970–99. The B1 scenario assumes substantial reductions in greenhouse gas emissions. The A2 scenario assumes continued increases in these emissions.

Figure 23.3 Richard Somerville teaching graduate students in 1989 at Scripps Institution of Oceanography, University of California, San Diego.

and there will always be more to learn. However, the great unknown about climate change in coming decades will not be some arcane and puzzling aspect of climate science that we do not yet understand well enough. The great unknown about climate change will be whether enough people will care enough, and soon enough, to take the actions needed to avoid severe climate damage.

For a scientist like me, recently retired from research, communicating climate change science to the broad public is the most important work I can do. In a sense, I am still teaching, but now I teach by writing books and articles and giving lectures for the public. I am no longer teaching graduate students in a classroom (Figure 23.3). I stress the urgency of mitigating climate change. Meeting the challenge of climate change successfully may be the single most important issue facing humanity. Failure to limit climate change to relatively safe levels can mean ultimately capitulating to an existential threat to human civilization.

In climate change, the entire world faces an urgent existential challenge. We are literally creating the planet our children will live in. We must act energetically and wisely. That is why climate science research and climate science communication are so crucial. Everything depends on what people and their governments do.

Everything depends on what people and their governments do.

Appendix: Curriculum Vitae of Richard C. J. Somerville

Richard C. J. Somerville

Distinguished Professor Emeritus
Scripps Institution of Oceanography
University of California, San Diego
https://rsomerville.scrippsprofiles.ucsd.edu
http://richardsomerville.com

Education

New York University (NYU), Meteorology, PhD, 1966
Pennsylvania State University, Meteorology, BS, 1961

Positions

2007–present: Distinguished Professor Emeritus, Scripps Institution of Oceanography, University of California, San Diego (UCSD)
2007–2019: Research Professor, Scripps Institution of Oceanography, UCSD
2004–2007: Distinguished Professor, Scripps Institution of Oceanography, UCSD
1979–2007: Professor, Scripps Institution of Oceanography, UCSD
1974–1979: Scientist and Section Head, National Center for Atmospheric Research
1971–1974: Research Scientist, Goddard Institute for Space Studies, National Aeronautics and Space Administration (NASA)
1971–1974: Adjunct Professor, Columbia University and NYU
1969–1971: Visiting Scientist, Courant Institute of Mathematical Sciences, NYU
1967–1969: Research Associate, Geophysical Fluid Dynamics Laboratory, National Oceanic and Atmospheric Administration (NOAA)
1966–1967: Fellow, Advanced Study Program, National Center for Atmospheric Research

Service

Coauthor, *Climate Mathematics: Theory and Applications*, Cambridge University Press, 2019

Member, Science and Security Board, *Bulletin of the Atomic Scientists*, 2012–2019

Climate Communication, Science Adviser, http://climatecommunication.org

Lecturer and popular author on climate change, http://richardsomerville.com

Climate Central, Member, Board of Directors, 2008–2011, www.climatecentral.org

Author, *The Forgiving Air: Understanding Environmental Change*, 2nd ed., 2008

Coordinating Lead Author, Intergovernmental Panel on Climate Change, www.ipcc.ch, 2007

Adviser and participant, Aspen Global Change Institute, 1989–2018, www.agci.org/

Coordinator, Climate Curricular Group, Scripps Institution of Oceanography, UCSD, 1995–1999

Director, Climate Research Division, Scripps Institution of Oceanography, UCSD, 1989–1996

Head, Climate Research Group, Scripps Institution of Oceanography, UCSD, 1979–1989

Past member of the Council and several committees of the American Meteorological Society

Trustee (1990–1995) and past Board Chair, University Corporation for Atmospheric Research

Advisory committee service: Department of Energy, NASA, NOAA, National Center for Atmospheric Research, National Academy of Sciences, National Research Council, National Science Foundation, University Corporation for Atmospheric Research

Honors

Haagen-Smit Clean Air Award, 2020–2021, of the California Air Resources Board, for a "lifetime of important contributions and achievements" in the field of climate change science

Spilhaus Ambassador Award Grant of the American Geophysical Union, 2018, supports work in societal impact, community service, scientific leadership, and promoting talent and the career pool

Ambassador Award of the American Geophysical Union, 2017, given for outstanding contributions in areas including societal impact, community service, scientific leadership, and promoting talent and the career pool

Fellow, American Geophysical Union, 2017, an honor given for "for exceptional contributions to Earth and space sciences"

Climate Communication Prize of the American Geophysical Union, 2015, given to one scientist annually, with a cash grant, for excellence in the communication of climate science

Edward A. Dickson Emeriti Professorship, UCSD, 2014, awarded to one outstanding UCSD emeritus professor each year, with a cash grant to support further work

Distinguished Alumni Award, Pennsylvania State University, 2011, the highest honor given by the university to an individual, usually given to fewer than 10 persons each year whose "personal lives, professional achievements, and community service exemplify the objectives of their Alma Mater"

Charles L. Hosler Alumni Scholar Medal, Pennsylvania State University, 2009, for alumni "who have made contributions to the development of science through research, teaching, or administrative leadership"

Lee Kuan Yew Distinguished Visitor, Singapore, 2007, "invites internationally eminent and outstanding academics and scholars to Singapore"

Coordinating Lead Author, Intergovernmental Panel on Climate Change, 2007, the title of heads of chapter writing teams of an IPCC assessment report. The IPCC shared the 2007 Nobel Peace Prize

Distinguished Professor, UCSD, 2004, the highest faculty rank, "is reserved for scholars and teachers of the highest distinction, whose work has been internationally recognized and acclaimed and whose teaching performance and service are excellent"

Distinguished Lecturer, Sigma Xi, The Scientific Research Society, 2002–2003, signifies being chosen to give a series of lectures at several universities

Louis J. Battan Author's Award, American Meteorological Society, 2000, for *The Forgiving Air: Understanding Environmental Science*. The award "is presented to the author of an outstanding, newly published book on atmospheric science"

Walter Orr Roberts Lectureship, American Meteorological Society, 1999, an award "presented for significant contributions to the understanding and discussion of global environmental change derived from multidisciplinary research activities"

Fellow, American Association for the Advancement of Science, 1993, "an honor bestowed upon members by their peers, for meritorious efforts to advance science or its applications"

Outstanding Contributions Award, San Diego Science Educator's Association, 1991, an award recognizing outstanding university science teachers

Fellow, American Meteorological Society, 1987, an award for those who "have made outstanding contributions to the atmospheric or related oceanic or hydrologic sciences or their applications"

Publications: https://rsomerville.scrippsprofiles.ucsd.edu/publications/

Glossary

In this book, some terms may be unfamiliar or technical. In numerous other books about climate change, encountering many more such terms may be unavoidable. This glossary is intended to be helpful in learning the vocabulary of atmospheric and climate science and related fields.

Absolute humidity is one way to measure atmospheric moisture content. It is the mass of water vapor per unit volume of the air containing the water vapor, typically expressed as grams of water vapor per cubic meter of air.

Absorption is the conversion into heat of a fraction of the radiation that is incident on an object.

Acid deposition is the combination of dry deposition plus precipitation (rain, snow, etc.) made acidic, typically by the addition of sulfur dioxide and nitrogen oxides into the atmosphere as a result of fossil fuel burning. Automobile exhaust and emissions from coal-fired power plants are two significant causes of acid deposition. In severe cases, acid deposition has rendered entire lakes and forests nearly lifeless.

Acid precipitation is rain (or snow) with a pH less than about 5.6, usually because of the presence of nitric and/or sulfuric acid. Natural rain is slightly acidic, so rain with a pH value somewhat less than the strictly neutral criterion (a pH of 7) is not generally referred to as acid rain.

Adiabatic means without the exchange of heat, in meteorology usually between an air parcel under consideration and its surroundings.

Advection in meteorology is horizontal movement of air or horizontal transport by air of any property, such as heat or humidity.

Aerosol is a gaseous suspension of fine liquid or solid particles in air.

Air mass is a large volume of air, typically thousands of kilometers in horizontal dimensions, that is relatively uniform horizontally in its properties, especially in temperature and moisture content.

Albedo is the fraction of light (or, more generally, of electromagnetic radiation flux) striking a surface that is reflected by that surface, often expressed as a percentage. Light colored surfaces, such as snow and ice, have a high albedo, while dark, light-absorbing surfaces have a low albedo.

Antarctic ozone hole is a severe depletion of ozone in the Antarctic stratosphere, due to the catalytic destruction of ozone by chlorofluorocarbons and related compounds.

Anthropogenic means human-caused or created by human beings. For example, anthropogenic carbon dioxide emissions are those caused by human activities such as burning fossil fuels.

Atmosphere is the envelope of gases that surrounds a planet.

Atmospheric pressure is the force exerted by the air on each unit of area of a surface, essentially equivalent to the weight of the overlying atmosphere.

Atom is the smallest unit of a chemical element that can take part in a chemical reaction. An atom is composed of a nucleus that contains protons and neutrons and is surrounded by electrons.

Barometer is an instrument used to measure atmospheric pressure.

Biodiversity is a general term that describes several aspects of the variety of the complex web of living things, and it has different specific meanings. Species diversity, the most common use of the term, refers to the total number of biological species in a particular area. Habitat diversity refers to the variety of places where life exists. Genetic diversity is primarily the variety of populations comprising each species, where population in this context refers to members of a species that live together and so can mate with one another.

Biosphere refers to the region on land, in the oceans, and in the atmosphere inhabited by living things.

Black body is an idealized or theoretical object that absorbs all of the radiation incident upon it and emits the maximum possible radiation for its temperature at every wavelength.

Business-as-usual, in the context of global climate change, refers to a scenario for future world patterns of energy use and greenhouse gas emission, which assumes that there will be no significant change in people's attitudes and priorities, so emissions will continue with little or no reductions.

Carbon cycle is the exchange of carbon between land, atmosphere, and oceans. About one quarter of the total quantity of atmospheric carbon (in the form of carbon dioxide) is cycled in and out of the atmosphere each year. Approximately half of this amount is exchanged with the land biota, and the other half, through physical and chemical processes, passes across the ocean surface.

Carbon dioxide (CO_2) is a colorless and odorless gas, which is a trace constituent of the Earth's atmosphere and is one of the major greenhouse gases. Human-caused CO_2 emissions result mainly from burning fossil fuels (coal, oil, and natural gas) and from deforestation and land use changes.

Celsius is a temperature scale, also called the Centigrade scale. Its fixed points are the freezing point of water (0°C) and the boiling point of water (100°C). To convert from Celsius to Fahrenheit, multiply the Celsius temperature by 1.8 and add 32°.

Chlorofluorocarbons (CFCs) are synthetic compounds invented by Thomas Midgley, Jr., in 1928 as refrigerants. They destroy stratospheric ozone and are also greenhouse gases. They have been used as a coolant in refrigerators and air conditioners. They have also been used as solvents, foam blowing agents, and aerosol propellants, though the use of CFCs in aerosol cans in the United States was outlawed in the late 1970s. Substitutes for CFCs are available. The Montreal Protocol for the Protection of the Ozone Layer is an international agreement that required developed nations to phase out CFCs.

Climate refers to the temperature, humidity, precipitation, winds, radiation, and other meteorological conditions characteristic of a locality or region over an extended period of time. Compared to weather, climate involves longer times and deals not only with the atmosphere, but also with oceans, land and biosphere. The essential characteristic of climate is that it is a statistical concept embracing the sum total of weather, thus including not only average conditions but also probabilities of extreme events and other descriptions of the variability of meteorological conditions.

Climate sensitivity refers to the magnitude of the climate change expected to result from a change in external influences. One frequently encountered description of climate sensitivity is the global average temperature rise expected to result from a doubling of carbon dioxide in the atmosphere.

Climatology is the scientific study of climate, particularly of its variability and its dependence on factors that influence its behavior.

Cloud is a visible mass of condensed water vapor (liquid or ice) particles in the atmosphere.

Concentration in climate science is a measure of the amount of a specific substance, such as a greenhouse gas, in the atmosphere. While emissions refer to the release of a gas into the atmosphere, concentration refers to the amount of a gas that is present in the atmosphere. Increased emissions generally lead over time to increased atmospheric concentrations. Concentrations are typically measured in parts per million (ppm), parts per billion (ppb), or parts per trillion (ppt). For example, a carbon dioxide (CO_2) concentration of 427 parts per million by volume (ppmv) means that there are 427 molecules of carbon dioxide in every million molecules of air. Increased concentrations of greenhouse gases such as carbon dioxide generally cause the greenhouse effect to become stronger.

Convection in meteorology generally refers to vertical motion in the atmosphere or ocean resulting in the transfer of heat.

Cyclone is a weather system characterized by relatively low atmospheric pressure compared with its surroundings, together with winds blowing counterclockwise in the Northern Hemisphere and clockwise in the Southern Hemisphere.

Deforestation is destruction of forests, usually by cutting or burning. Deforestation increases the greenhouse effect in two ways. First, when wood is burned or decomposes, it releases carbon dioxide. Second, trees that are destroyed can no longer remove carbon dioxide from the atmosphere by photosynthesis.

Drought is an extended period of abnormal dryness for a particular region.

Ecosystem refers to a distinct system of interdependent plants and animals, together with their physical environment. An ecosystem may be as large as the entire Earth, or as small as a pond.

El Niño and La Niña are climate patterns that can affect weather in many regions. El Niño is a warming of the surface waters of the eastern tropical Pacific Ocean, occurring at irregular intervals of about two to seven years, usually lasting from several months up to about two years, which has a significant influence on regional and global climate. El Niño has been linked to colder, wetter winters in parts of the US, drier hotter summers in South America and Europe, and drought in Africa, as well as reduced numbers of fish in South American Pacific coastal waters due to reduced upwelling of nutrient-rich waters. La Niña is an opposite phase of this pattern and is sometimes referred to as simply a cold event. Niño and Niña both mean child. They are, respectively, the masculine (boy) and feminine (girl) forms of the word in Spanish.

El Niño–Southern Oscillation (ENSO) refers to the closely linked phenomena of El Niño (see the previous glossary term) and a global-scale shift in atmospheric pressure called the Southern Oscillation (SO). In the so-called warm phase of ENSO, El Niño warming extends over much of the tropical Pacific and becomes clearly linked to the SO pattern. Many of the countries most affected by ENSO events are developing countries with economies that are largely dependent upon their agricultural and fishery sectors as a major source of food supply, employment and foreign exchange. New capabilities to predict the onset of ENSO events can thus have important human impacts. ENSO is fundamentally an aspect of the combined ocean-atmosphere system and cannot be understood as either entirely meteorological or entirely oceanographic in nature. While ENSO is a natural part of the Earth's climate variability, whether its intensity or frequency may change as a result of global warming is a concern.

Electromagnetic radiation is energy transfer by waves with both electrical and magnetic properties.

Electromagnetic spectrum is a range of types of radiation, ordered by wavelength or frequency, including all kinds of electric, magnetic, and visible radiation.

Electron is a negatively charged component of an atom.

Element refers to any substance that cannot be separated by chemical means into two or more simpler substances.

Emissions in climate science are substances released into the atmosphere, specifically greenhouse gases such as carbon dioxide that absorb infrared radiation and cause warming.

Environment is the complex of physical, chemical, and biological factors in which a living organism or community exists.

Environmental refugees are people obliged to leave their traditional or established homelands due to environmental problems (such as deforestation, desertification, floods, drought, and sea-level rise), on a permanent or semipermanent basis, sometimes with little or no hope of returning.

Equator is an imaginary circle around the Earth that is equally distant from the North and South Poles and defines the latitude 0°.

Evaporation is the process of changing state from liquid to vapor (gas).

Fahrenheit is a temperature scale based on water freezing at 32°F and boiling at 212°F under standard atmospheric pressure. To convert from Fahrenheit to Celsius, subtract 32° from the Fahrenheit temperature and divide the resulting quantity by 1.8.

Feedback refers to a sequence of interactions in which the final interaction influences the original one. In such a sequence, a cause produces a result, and the result then in turn influences its cause. As a system changes, it may generate processes that affect the original change. If one of these processes amplifies the change, it is called a positive feedback. If it reduces the change, it is called a negative feedback.

Food chain refers to a series of plants and animals that depend upon each other as food sources (i.e., a plant is eaten by a small fish, which is eaten by a larger fish, which is eaten by a bird, and so on).

Fossil is the hardened remains or traces of plant or animal life from a previous geological period preserved in the Earth's crust.

Fossil fuels such as coal, oil, and natural gas are created by the decomposition of ancient animal and plant remains. They are finite (limited) resources, and they release carbon dioxide and other gases when burned.

Front is a narrow zone marking the boundary between two air masses of significantly different meteorological properties, usually of temperature, humidity, and wind speed and direction.

General circulation models (GCMs) are computer models of the Earth's climate that are used to improve our understanding of factors that influence climate and enhance our ability to forecast future climate patterns. One reason GCMs are so useful is that they allow researchers to vary individual factors and observe the results, isolating processes in a way that is not possible in the physical world. GCMs are also sometimes called global climate models, a more accurate and descriptive characterization. "General circulation" is an older technical term in meteorology, which originally referred to the long-term average aspects of large-scale atmospheric behavior. In modern usage, a GCM can be an atmospheric model or an oceanic one or a model of the coupled climate system involving the atmosphere and ocean and other components as well.

Geoengineering refers to intentional artificial modification of the Earth systems intended to counteract human-caused effects such as global warming or stratospheric ozone depletion. Some geoengineering proposals would be costly and logistically difficult. Before geoengineering is undertaken, it would be prudent to be certain that unanticipated adverse consequences would not occur. The management of geoengineering involves many uncertainties.

Glacier is a multiyear accumulation of snowfall in excess of snow melt on land, resulting in a mass of ice covering at least a tenth of a square kilometer, that shows some evidence of movement in response to gravity. Glacier ice is the largest reservoir of fresh water on the Earth and second only to the oceans as the largest reservoir of total water. Glaciers are found on every continent except Australia.

Global change refers to changes in the Earth's system, which is either a global phenomenon or that occurs regionally, but strongly enough and often enough to be of global significance. Leading current global change issues include climate change due to a strengthened greenhouse effect, stratospheric ozone depletion, acid precipitation, urban air pollution, and loss of biodiversity.

Global warming is a term used to describe a warming of the Earth's climate due to increased concentrations of greenhouse gases in the atmosphere.

Greenhouse effect is the natural process whereby gases in the Earth's atmosphere act like the glass in an idealized greenhouse, letting the Sun's energy in, but preventing some of the Earth's radiation from escaping to space. Were it not for this natural effect, the Earth's climate would be about 33°C (60°F) colder, and life as we know it would not exist. The "increased greenhouse effect" refers to an increase in this natural heat-trapping phenomena due to human-caused emissions of greenhouse gases. Actual greenhouses made of glass can also reduce heat loss because the glass walls and roof prevent the wind from carrying away heat. This process is different from the greenhouse effect.

Greenhouse gases are water vapor, carbon dioxide, methane, tropospheric ozone, nitrous oxide, CFCs, and other gases which absorb and re-radiate some of the long-wave (infrared) radiation emitted from the Earth's surface, thereby contributing to the greenhouse effect and warming the atmosphere. With the exception of water vapor, these gases are also called "trace gases" because in total they make up less than 1% of the atmosphere.

Hail is precipitation composed of lumps of ice. Hail is produced when large frozen raindrops, or other particles in cumulonimbus clouds, grow by accumulating super-cooled liquid droplets. Strong updrafts in the cloud carry the particles up through the freezing air, allowing the frozen core to accumulate more ice. When a piece of hail becomes too heavy to be carried by rising air currents, it may fall to the ground.

Humidity is the amount of water vapor in the air. The higher the temperature, the greater the amount of water vapor that the air can hold. "Relative humidity" describes the amount of water in the air compared with the maximum amount of water that the air could hold at the same temperature and pressure. For example, 50% relative humidity means the air holds half of the amount of water vapor that it is capable of holding.

Hurricane is an intense warm-core tropical storm with winds exceeding 74 miles per hour. Hurricanes typically originate over regions such as the tropical and subtropical North Atlantic and North Pacific oceans, because high sea surface temperatures are essential to their formation.

Hydrologic cycle refers to the natural sequence through which water evaporates from the ocean, land surface, and plants into the atmosphere as water vapor, falls to the Earth as precipitation, and largely returns to the ocean through pathways including rivers and ground water.

Ice Age is a geological time period during which sheets of ice cover extensive parts of the Earth.

Indigenous means naturally occurring in an area, such as an indigenous species.

Infrared radiation is electromagnetic radiation at wavelengths longer than the wavelengths of red visible light, but shorter than microwaves. Most radiation emitted by the Earth is infrared, and it is this radiation which is involved in the greenhouse effect.

Intergovernmental Panel on Climate Change is the United Nations body for assessing the science related to climate change. This organization, often referred to as IPCC, involves many of the world's leading scientific authorities on climate change. IPCC assessment reports, published at intervals of several years, are among the most authoritative summaries of scientific knowledge of climate change. The IPCC is mandated to be policy-relevant but not policy-prescriptive.

Isothermal means of or indicating equality of temperature. In meteorology, isotherms are lines connecting points of equal temperature on a weather map.

Kilometer is a metric unit of distance approximately equal to 3,280.8 feet or 0.621 miles.

Lightning is a very strong discharge of atmospheric electricity accompanied by a vivid flash of light. Electric charges build up through complicated processes. Air initially acts as an insulator between opposite charges. When charges are strong enough, this insulating property of air can break down and lightning can occur in clouds, between clouds, and between clouds and ground.

Mean is the scientific term for arithmetic average, as in "global mean temperature."

Meteorology is the science of the atmosphere.

Mitigation is the action of reducing the severity, seriousness, or painfulness of something. When applied to climate change, mitigation often refers to reducing or preventing emissions of greenhouse gases and other heat-trapping substances.

Modeling is an investigative technique that uses a mathematical or physical representation of a system or theory to test for effects that changes in system components may have on the overall functioning of the system. Mathematical modeling using computers plays a major role in scientific research on climate change, by simulating how the Earth's climate will respond to changes in atmospheric concentrations of greenhouse gases.

Molecule means two or more atoms of one or more elements chemically combined in fixed proportions. For example, atoms of the elements carbon and oxygen, chemically bonded in a 1:2 proportion, create molecules of the compound carbon dioxide (CO_2). Molecules can also be formed of a single element, as in ozone (O_3).

Monsoon refers to a particular seasonal weather pattern in subtropical regions, especially when characterized by periods of heavy winds and rainfall. Monsoons are caused by a pronounced seasonal change in wind direction. Winds usually blow from land to sea in winter, while in the summer, this reverses, bringing precipitation. Monsoons are most typical in India and southern Asia.

Montreal Protocol on Substances that Deplete the Ozone Layer is an international agreement that prescribed a timetable for ending the production of chlorofluorocarbons (CFCs) and related compounds. Begun in 1987, this unprecedented international treaty is a unique example of scientists and industry working with governments to seek a global solution to the human-caused environmental challenge of ozone depletion. After the original agreement was signed, new evidence arose proving that deeper cuts in CFC production were necessary to protect the ozone layer. The 1990 London amendments and the 1992 Copenhagen amendments sped up the halocarbon phaseout and controlled several other chemicals that were not in the original agreement: methyl chloroform, carbon tetrachloride, methyl bromide, and HCFCs. The treaty also attempted to make the phaseouts fair to developing countries by setting up a fund, paid for by developed nations, to assist developing countries in making the switch to ozone-safe chemicals.

Orbit is the path of a body, such as a planet or satellite, in its periodic revolution around another body in space. For example, satellites that orbit the Earth near latitude 0° are said to have equatorial orbits, since they remain above the equator. Satellites with inclinations near 90° are said to be in polar orbits because they cross over or near the Earth's north and south poles as they revolve around the planet.

Ozone is a gaseous molecule consisting of three atoms of oxygen (O_3). Ozone in the Earth's stratosphere forms a protective layer that shields the Earth's inhabitants from damaging ultraviolet radiation from the Sun. Ozone in the troposphere, near the Earth's surface, on the other hand, is a harmful pollutant resulting from the interaction of anthropogenic emissions of nitrogen oxides and volatile organic compounds and sunlight.

Ozone depletion refers to the thinning of the stratospheric ozone layer which protects life on the Earth from excess ultraviolet radiation from the sun. Human-made halocarbons are primarily responsible for this reduction in the amount of ozone in the stratosphere.

Ozone hole refers to a region of the atmosphere over Antarctica where, during springtime in the Southern Hemisphere, a substantial fraction of the stratospheric ozone disappears. Halocarbons produced by human activities, such as chlorofluorocarbons, are the primary cause of this phenomenon.

pH is a measure of the acidity or alkalinity of a solution. A value of 7 is neutral, values less than 7 are acidic, and values over 7 are alkaline or basic. A change of 1 unit on the pH scale represents a factor of 10 in acidity. For example, a solution with a pH of 5 is 10 times as acidic as one with a pH of 6.

Photosynthesis is the series of chemical reactions by which plants use the sun's energy, carbon dioxide, and water vapor to form materials for growth, and release oxygen.

ppmv stands for parts per million by volume, a measure of concentration.

Precipitation in meteorology refers to liquid or solid forms of moisture that fall from clouds, including rain, snow, and hail. Raindrops typically form around condensation nuclei, which are often particles of salt or dust. Under appropriate conditions, water or ice droplets forming on these particles can attract more water and continue to grow until they are large enough to fall out of the cloud.

Rainforest is an evergreen woodland of the tropics distinguished by a continuous leaf canopy and an average rainfall of about 100 inches (250 centimeters) per year. Rainforests play an important role in the global environment for several reasons. They are the most biologically diverse biome on the planet, encompassing just 6–7% of the Earth's land, but thought to house nearly half of its species. Rainforests also take up carbon dioxide, helping to balance human-caused emissions. When rainforests or other forests are cut or burned, the opposite occurs: They release stored carbon dioxide, adding to the greenhouse effect.

Remote sensing is the process of obtaining information from a distance, especially from aircraft and satellites. Modern remote sensing technology has greatly expanded our ability to see and understand the Earth and its systems and to observe changes. Remote sensing has become a critical tool in activities ranging from the verification of arms control treaties to the provision of emergency aid to disaster-stricken regions. Through remote sensing, we learn about problems such as droughts, famines, and floods; we obtain information about agricultural practices, weather conditions, transportation systems, river flows, and terrain changes. We use remote sensing to locate the Earth's natural resources and can then use that information to exploit or protect them.

Renewable energy refers to energy from sources that are not depleted by use. Examples include using passive solar energy to heat buildings, solar thermal energy to heat water or turn turbines to produce electricity, and photovoltaic cells to convert sunlight directly to electricity, wind power, and hydroelectric energy.

Sequestration is removal and storage, as when carbon dioxide is sequestered from the atmosphere by plants via photosynthesis. An important research issue in climate change mitigation now is concerned with finding economical and practical ways to capture and store the carbon dioxide produced by burning fossil fuels, thus reducing the emission of carbon dioxide to the atmosphere.

Solar radiation is energy from the sun. It is the main energy source for the Earth's climate system, heating the surface of the Earth and driving currents in the oceans and winds in the atmosphere. Ordinary visible sunlight is the most obvious form of solar radiation, but other forms are significant too. For example, see ultraviolet radiation.

Stratosphere is the region of the atmosphere between the troposphere and mesosphere, having a lower boundary approximately 8 kilometers above sea level at the poles and 15 km at the equator and an upper boundary at an altitude of approximately 50 km above sea level. This is the region that contains the ozone layer which protects the Earth' surface from ultraviolet solar radiation.

Sustainable development is development that meets the needs of the present generation without compromising the ability of future generations to meet their own needs. Some people also think that the concept of sustainable development should include preserving the environment for other species as well as for people.

Terrestrial means pertaining to the Earth, as distinct from other planets (as in extraterrestrial life). It also means pertaining to the land, as distinct from the water or air (as in a terrestrial, as opposed to aquatic, ecosystem).

Thunder is the sound that results from lightning. A lightning bolt produces an intense burst of heat which makes the air around it expand explosively, producing the sound we hear as thunder. Since light travels much faster than sound, we see the lightning before we hear the thunder. The difference in time between the two can tell us how far away the clouds producing the lightning and thunder are.

Thunderstorm is a local storm resulting from rising warm humid air, which produces lightning and therefore thunder, usually accompanied by rain or hail, gusty winds, and strong vertical air motion.

Tornado is a strong, rotating column of air extending from the base of a cumulonimbus cloud to the ground. These twisting, spinning funnels of low-pressure air are a violent weather event, capable of causing great damage. They are created during powerful thunderstorms.

Tropics refers to the region of the Earth from latitude 23.5° north (the Tropic of Cancer) southward across the equator to latitude 23.5° south (the Tropic of Capricorn). This region has relatively small daily and seasonal changes in temperature, but great seasonal changes in precipitation.

Ultraviolet radiation (UV) is the energy range just beyond the violet end of the visible spectrum. Most UV is blocked by the Earth's atmosphere (particularly the stratospheric ozone layer), but some solar UV penetrates and aids in plant photosynthesis and the production of Vitamin D in humans. Too much UV can burn the skin, cause skin cancer and cataracts, and damage vegetation.

Volcano is a naturally occurring vent or fissure at the Earth's surface through which erupt molten, solid, and gaseous materials. Volcanic eruptions inject large quantities of dust, gas, and aerosols into the atmosphere and can thus cause temporary climatic cooling.

Water vapor is the invisible, gaseous form of water.

Weather is the state of the atmosphere at some place and time, particularly as characterized by variables such as temperature, cloudiness, wind, humidity, and precipitation.

Wind is a natural motion of the air, especially a noticeable current of air moving in the atmosphere approximately parallel to the Earth's surface.

Resources: Recommended Websites and Books

Recommended Websites

I highly recommend these websites:

https://climatecommunication.org – the site operated by Susan Joy Hassol, all about communicating climate change science.

https://climate.nasa.gov – the climate site of NASA, the US space agency, with some beautiful visuals and lots of climate news.

https://skepticalscience.com – a wonderful site for refuting the claims of people who, often for mysterious reasons of their own, do not accept the results of mainstream climate science.

www.ipcc.ch – the site of the IPCC, the Intergovernmental Panel on Climate Change, on which are many downloadable reports. Warning: Abundant mathematics and jargon await you here.

Recommended Books

The subject of climate change has an immense literature. Here is a list of several varied and recently published books. I have not tried to select the "best books," which in any case would be a different list for different readers. I am confident that most readers will be able to find several books on this list that will be interesting and informative for them.

Alley, Richard B. *Earth – The Operator's Manual*. W. W. Norton & Company, 2011. This book is slightly out of date, but it is an engagingly written account of how we came to learn that carbon dioxide has built up in the atmosphere and will cause damaging climate change. Alley is an outstanding scientist and is renowned as an energetic speaker and an effective communicator. The book contains no mathematics and assumes no scientific background on the part of the reader. In fact, it was written to accompany a PBS television documentary. Alley writes in much the same light-hearted and warm tone that he uses when speaking.

Bennett, Jeffrey. *A Global Warming Primer: Answering Your Questions about the Science, the Consequences, and the Solutions*. Big Kid Science, 2024. This essentially self-published book is attractive in several ways. It is chatty and optimistic, promising a future "in which today's children will someday be able to talk about global warming as a once-serious problem that we found a way to solve." I contributed a favorable blurb to it: "From science to solutions, this clearly written and up-to-date survey of human-caused climate change illuminates one of the great existential issues of our time." – Richard C. J. Somerville. However, I do not agree with everything in this book. For example, the author contrasts climate skeptics versus climate believers. This framing in my view is a false dichotomy. Science is based on facts and evidence. "Everybody is entitled to their own opinions, but not to their own facts." Facts are objective truths. I do not "believe" the Earth orbits around the sun or rotates on its axis. Those are facts. I do "believe" people should be kind and honest. That is an opinion.

Bolin, Bert. *A History of the Science and Politics of Climate Change: The Role of the Intergovernmental Panel on Climate Change*. Cambridge University Press, 2007. This book is both a scholarly work of history and a scientific autobiography. The book, like Bert Bolin, is a model of clear communication, modesty, optimism and vision. Bolin was thrilled when the announcement came in October 2007 that the 2007 Nobel Peace Prize was to be awarded in equal shares to Al Gore and to the Intergovernmental Panel on Climate Change (IPCC). He was too ill to attend the Nobel ceremony in Oslo in December 2007, and he died in Stockholm on December 30, but Gore and many others emphasized the key role that Bolin had played in the success of the IPCC. Bolin's legacy of intellectual integrity and scientific leadership is embodied in the reports of the IPCC. In fact, the IPCC exemplifies the ideal of sound science in the service of society, a concept in which Bert Bolin believed deeply and which guided his life. Bolin's own scientific eminence was a key reason for the willingness of many leading experts to contribute to the work of the IPCC. Today, the organization enjoys a reputation for fairness, transparency, and inclusiveness in its work, and this too continues a tradition established under Bolin. The IPCC itself is a lasting memorial to Bert Bolin. Wren's famous epitaph is apt: "Lector, si monumentum requiris, circumspice." "Reader, if you seek his monument, look around you."

Emanuel, Kerry A. *Climate Science, Risk & Solutions*. Massachusetts Institute of Technology, 22 March 2024: https://climateprimer.mit.edu/climate-risk-solutions.pdf. Kerry Emanuel is Professor Emeritus of Atmospheric Science at MIT. This up-to-date little book is very clearly written, is about 50 pages long, contains about 20 exceptionally informative graphs, and can easily be downloaded from the above MIT website at no cost. As an added benefit, it includes a questions-and-answers chapter devoted to dispelling climate myths. This chapter is written by Susan Joy Hassol, an expert in communicating climate information, who has also collaborated extensively with me, as I have described earlier. Professor Emanuel's "climate primer" is extremely well-written, and it includes a valuable discussion of how society must now evaluate risks and opportunities and recognize the need to create a better world for the future. I highly recommend this book. I urge everyone to read it. Kerry Emanuel has told me that he credits Susan Joy Hassol and "a team of gifted and enthusiastic web designers at MIT" for developing the award-winning website on which anyone can read and download this book for free. I credit him for writing an excellent book and making it available on an excellent website for free.

Emanuel, Kerry A. *What We Know about Climate Change*, updated edition. The MIT Press, 2018. This valuable little book is very readable. It uses no mathematics and contains only two simple graphs. It is only about 70 pages long, and that number includes about 20 blank pages between chapters. Yet it manages to convey a great deal of accurate and authoritative information. The world needs such a book. Unusually for a climate change book aimed at a general audience, this one begins by introducing the reader to paleoclimatology, the study of ancient climates in the distant geological past. Lacking data from modern instrumental records, paleoclimatology largely relies on proxy data such as tree rings and ice cores. Emanuel shows how learning about the history of climate change can add to our understanding of climate variability. As he writes, paleoclimatology "has produced among the most profound yet least celebrated scientific advances of our era."

Henson, Robert. *The Rough Guide to Climate Change*, third edition. Rough Guides, 2011. I contributed a favorable blurb to an earlier edition of this book: "Scientifically up-to-date and clearly written, this courageous book cuts through mystery and controversy to explain climate change for readers who prefer facts." – Richard C. J. Somerville. The author is a skillful science writer. He is well informed and worked for years at the National Center for Atmospheric Research in Boulder, Colorado.

Kiehl, Jeffrey T. *Facing Climate Change: An Integrated Path to the Future*. Columbia University Press, 2016. Jeffrey Kiehl is a distinguished climate scientist who is also a Jungian psychoanalyst. It is safe to say that this is a rare combination of careers and expertise. His book is unique too. He describes the task of meeting the threat of climate change by drawing on the perspectives and wisdom of both climate science and Jungian psychotherapy.

Klein, Naomi. *This Changes Everything*: *Capitalism vs The Climate*. Simon & Schuster, 2014. A well-known and best-selling Canadian author and public intellectual, Naomi Klein in this book argues that successfully dealing with the existential issue of climate change can best be accomplished by making major changes in the dominant economic system of the world today.

Kolbert, Elizabeth. *Field Notes from a Catastrophe*. Bloomsbury Publishing, 2006. Kolbert is a staff writer for *The New Yorker* magazine, where the material in this book was first serialized. She brings to climate change a penetrating intellect and the talents of an exceptional writer. The science she relates is accurate and well-described, but the great strength of this book is its treatment of real people, both those doing climate research and those affected by climate change. When people ask me to recommend just one book on climate change, one that is brief and readable and trustworthy and not laden with equations or jargon, this is the one.

I published these words in 2008 in my book, *The Forgiving Air: Understanding Environmental Change*. American Meteorological Society. Now, in 2025, I would not change a word, except to add that the following two later books by Elizabeth Kolbert are also extraordinary, and anything she writes is certain to be well worth reading.

Kolbert, Elizabeth. *The Sixth Extinction: An Unnatural History*. Henry Holt, 2014. This is the book that won Elizabeth Kolbert a Pulitzer Prize. As she writes in her brutally honest and frank style, "Atmospheric warming, ocean warming, ocean acidification, sea-level rise, deglaciation, desertification, eutrophication – these are just some of the by-products of our species's success." The title comes from the five great mass extinctions in the past, each of which devastated the diversity of life on our planet. One such event was the violent impact of a large asteroid some 66 million years ago, which led to the extinction of the dinosaurs, except the birds. Thus, the sixth extinction is the one now occurring, and this one is entirely our fault. It is caused by the many devastating effects of human actions on the world of living things.

Kolbert, Elizabeth. *Under a White Sky: The Nature of the Future*. Crown, 2021. The title of this book comes from what is known as "solar radiation management," which is a form of geoengineering. The idea is that mankind might undertake intentional modification of the climate system, with the hope of countering some of the inadvertent modification that has led to harmful, human-caused climate change. One geoengineering approach might involve adding reflective particles to the atmosphere, which could lead to some cooling. There are many risks to such schemes, and many possible unanticipated consequences. After all, reflective particles are not simple negative greenhouse gas molecules. One known effect of solar radiation management is that the reflective particles could cause the sky to change color and become milky white and no longer blue. As just one citizen of the planet, my own view of geoengineering is similar to my view of nuclear war: study it, in order to better understand it, but never do it.

Mann, Michael E. *The Hockey Stick and the Climate Wars: Dispatches from the Front Lines*. Columbia University Press, 2012. Michael Mann is a prominent university professor and an outstanding climate scientist. Like many mainstream climate scientists, he has been the target of often vicious attacks by people variously described as climate contrarians or climate denialists. These are people who simply do not accept the conclusions reached by about 97% of climate experts. Here, the number 97% is based on substantial research. Professor Mann is a scientist who has chosen to fight back energetically against these people, using lawsuits and op-ed columns and congressional testimony and books such as this one. This book of more than 400 pages is the detailed story of Mann's continuing battles with his opponents. It is an excellent account of the politics and policies and battles of the partisan and contentious "climate wars." Mann is a prolific author, and I can enthusiastically recommend this book, especially to those wishing to learn about the battle between mainstream climate change scientists and those who denigrate and oppose them.

Mann, Michael E. *The New Climate War: The Fight to Take Back Our Planet*. Public Affairs, 2021. Michael Mann is a climate scientist who has written several books for the general reader, and all of them are well worth reading. This book is an effort to familiarize readers with Mann's struggle against deception, deflection, delay, denial, and disinformation carried out by certain fossil fuel interests and their supporters. Early in the book, he explains its purpose: "In the process of defending myself and my work from politically motivated attacks, I became a reluctant and involuntary combatant in the climate wars. I've seen the enemy up close, in battle, for two decades now. I know how it operates and what tactics it uses." Mann begins the book with evidence that scientists within the fossil fuel corporation ExxonMobil knew decades ago of the dangers of emitting large amounts of carbon dioxide into the atmosphere.

Nordhaus, William. *The Climate Casino: Risk, Uncertainty and Economics for a Warming World*. Yale University Press, 2013. William Nordhaus is a professor at Yale and an economist who has worked extensively on economic issues connected with climate change. He shared the 2018 Nobel Prize in Economic Sciences. The Nobel Foundation described Nordhaus's research in these words: "William Nordhaus's findings deal with interactions between society, the economy and climate change. In the mid-1990s, he created a quantitative model that describes the global interplay between the economy and the climate. Nordhaus's model is used to examine the consequences of climate policy interventions, for example carbon taxes." Furthermore, the Nobel Prize announcement included the statement that Nordhaus had "significantly broadened the scope of economic analysis by constructing models that explain how the market economy interacts with nature."

North, Gerald R. *The Rise of Climate Science: A Memoir*. Texas A & M University Press, 2020. My own appraisal of this book is in a blurb by me on the back of the dust cover: "A wonderful first-person scientific memoir, chronicling the development of modern climate science, written by a distinguished climate scientist who has contributed importantly to the story he tells. His perceptive and generous descriptions of many of his fellow scientists are a highlight of this very reasonable book." – Richard C. J. Somerville. This book is quite clearly conceived as an autobiography. The author tells the story of his life in about 300 pages. The book contains many details of his education and career. It includes many photographs, extensive accounts of his travels and scientific conferences, and anecdotal accounts of his interactions with numerous people. North is modest and self-deprecating throughout this book.

Saravanan, R. *The Climate Demon: Past, Present, and Future of Climate Prediction*. Cambridge University Press, 2022. I have written a blurb which is on the dust jacket and also on one of the first pages of the book: "A wide-ranging guided tour of the modern science of climate prediction, told by a leading expert without jargon or mathematics, and illuminated by history, philosophy, technology and even literature." – Richard C. J. Somerville. This book is primarily a description of climate science viewed in historical perspective. It could be used as a textbook for a course on the history of climate science. It does not require the reader to have any detailed capability in mathematics or science.

Stoknes, Per Espen. *What We Think about When We Try Not To Think about Global Warming: Toward a New Psychology of Climate Action*. Chelsea Green Publishing, 2015. A readable book by an unusual thinker. Stoknes is a Norwegian psychologist and politician with a doctoral degree in economic theory from the University of Oslo. He describes novel approaches to the challenge of confronting climate change.

Thunberg, Greta. *The Climate Book*. Penguin Press, 2023. Greta Thunberg, the well-known Swedish environmental activist, is superb and admirable in many ways. She is invariably modest and has frequently been careful to point out that she is not a scientist. I like *The Climate Book*, but I must point out that it is some 460 pages long and consists mainly of more than 100 short articles, all by different authors. These articles are devoted to a wide range of aspects of climate change, including policies and politics and many other topics, in addition to the science of climate change. However, Thunberg's volume may be rather daunting for some students, and indeed it does not speak with one voice. How could it possibly speak with one voice with so many authors writing on so many different topics? In my professional opinion, a few of the available good books on climate change have been written by science writers, celebrities, and others, including Bill Gates, Al Gore, and Greta Thunberg. This is one of the best of these good books.

References

Archer, David, and Raymond Pierrehumbert. *The Warming Papers: The Scientific Foundation for the Climate Forecast.* Wiley-Blackwell, 2011.

Arrhenius, Svante. On the influence of carbonic acid in the air upon the temperature of the ground. *Lond. Edinb. Dubl. Phil. Mag. J. Sci.*, 5th series, **41**(251): 39, 1896.

Arrhenius, Svante. *Worlds in the Making: The Evolution of the Universe.* Harper & Brothers Publishers, 1908.

Brooks, C. E. P. Geological and historical aspects of climate change. In *Compendium of Meteorology*, T. F. Malone, ed., American Meteorological Society, 1004–1018, 1951.

Crawford, Elisabeth. *Arrhenius: From Ionic Theory to the Greenhouse Effect.* Watson Publishing International, 1996.

Einstein, Albert. *Ideas and Opinions.* Crown Publishers, 1954.

Gladstone, John Hall. *Michael Faraday.* Macmillan & Co., 1873.

Intergovernmental Panel on Climate Change (IPCC). *Climate Change 2021: The Physical Science Basis.* Working Group I Contribution to the Sixth Assessment Report of the Intergovernmental Panel on Climate Change. Cambridge University Press, 2021.

Jackson, Roland. *The Ascent of John Tyndall: Victorian Scientist, Mountaineer, and Public Intellectual.* Oxford University Press, 2018.

Keeling, Charles D. Is carbon dioxide from fossil fuel changing man's environment? *Proc. Am. Philos. Soc.*, **114**: 10–17, 1970.

Keeling, Charles D. Rewards and penalties of monitoring the Earth. *Annu. Rev. Energy Environ.*, **23**: 25–82, 1998.

Oreskes, Naomi, and Erik M. Conway. *Merchants of Doubt: How a Handful of Scientists Obscured the Truth on Issues from Tobacco Smoke to Global Warming.* Bloomsbury Press, 2010.

Ramanathan, Veerabhadran, et al. *Bending the Curve: Climate Change Solutions.* Regents of the University of California. Retrieved from https://btc.ucsd.edu/, 2019.

Revelle, Roger, and Hans E. Suess. Carbon dioxide exchange between atmosphere and ocean and the question of an increase of atmospheric CO_2 during the past decades. *Tellus*, **9**: 18–27, 1957.

Shen, Samuel S. P., and Richard C. J. Somerville. *Climate Mathematics: Theory and Applications.* Cambridge University Press, 2019.

Siegenthaler, Ulrich, and Hans Oeschger. Predicting future atmospheric carbon dioxide levels. *Science*, **199**: 388–395, 1978.

Smil, Vaclav. *How the Earth Really Works: A Scientist's Guide to Our Past, Present and Future.* Viking, Penguin Random House, 2022.

Somerville, Richard C. J. *The Forgiving Air: Understanding Environmental Change*, 2nd ed. American Meteorological Society, 2008.

Somerville, Richard C. J. If I were President: A climate change speech. *Bull. Amer. Meteor. Soc.*, **89**: 1180–1182, 2008.

Somerville, Richard C. J. Medical metaphors for climate issues. *Clim. Change*, **76**: 1–6, 2006.

Somerville, Richard C. J., and Susan Joy Hassol. Communicating the science of climate change. *Phys. Today*, **64**: 10, 48–53, 2011.

Tyndall, John. The Bakerian Lecture: On the absorption and radiation of heat by gases and vapours, and on the physical connexion of radiation, absorption, and conduction. *Phil. Trans. Roy. Soc. Lond.*, **151**: 1–36, 1861.

World Meteorological Organization. State of the Global Climate 2024 (WMO-No. 1368), 2025.

Index

12 points
  summarizing fundamental findings of climate change science, 29–31, 58

advection, 106
aerosols, 10, 17, 94, 101, 145
albedo, 83, 84, 120–122, 137, 139–141
American Meteorological Society, 48, 75, 149
Antarctica, 6, 31, 36, 75, 93, 96, 137
  and sea level rise, 35, 116–117
Argo, 34–35, 96
argon, 92
Arrhenius, Svante, 47, 80, 81, 86, 88, 100–101, 103
Aspen Global Change Institute (AGCI), 46
atmosphere
  composition of, xii, xiii, 3, 13, 16, 31, 36–37, 92
  warming of, 17
Austen, Jane, 48
Australia
  and the Global Weather Experiment, 129

Balzac, Honoré de, 48
Bending the Curve: Climate Change Solutions, 53
biosphere, 91–92, 94
black body, 120–121
black carbon, 17, 94
boundary-value problem, 133
Brooks, C. E. P., 48
Bulletin of the Atomic Scientists, 1, 23, 156
Bunsen, Robert, 79

calculus, 104
carbon cycle, xiii, 94

carbon dioxide, xi
  annual global production of, 92
  concentrations, measurement of, xiii
  doubling preindustrial concentration of, 64, 81, 100, 109
  and the greenhouse effect, 77–78, 84–85
  historical record of atmospheric concentrations, 61
  increase in as international problem, xii, 18, 27, 91
  radiative signature of, 93
  rate of increase in, 17
  and removal from the atmosphere, 13, 18, 37
  sequestration of, 87
cement manufacture, 91
China, 6, 99
chlorofluorocarbons, 15, 87, 94, 100
climate feedbacks, 141
climate models, xi, 34, 103
  challenges of, 111, 113–114, 141
  and clouds, 140
  for doubled carbon dioxide level, 137
  and effects of oceans, 110
  reliability of, 136, 138
climate myths, 22–24, 38
climate sensitivity, 100, 136, 144–145
clouds
  average cloud cover of Earth, 121
  in climate modeling, 122
  effect on reflectivity, 118, 120, 141
  and greenhouse effect, 83, 119–120, 141
  and terrestrial radiation, 119
Cold War, xi
Compendium of Meteorology, 47
computational mathematics, 104
computers, xi, 104, 107, 110

Conference of the Parties (COP), 3
Copenhagen Accord, 16
Cruz, Ted, 21

Dalai Lama, 56, 158
Darwin, Charles, 80
data
  observational, uncertainty of, 96–97
deforestation, 91, 133, 150
Democratic party, 3, 40, 153
droughts, 101, 112, 144
Dust Bowl, 102

Earth Summit, 3, 15
eclipses
  prediction of, 104, 123
Einstein, Albert, xv, 1, 105, 108
El Niño, 26, 133
electromagnetic radiation, xv, 117–118
emissions, xi, 3, 15
energy, renewable, 50, 71, 157
energy supply, global, 86
European Centre for Medium-Range Weather Forecasts (ECMWF), 124
European Union, 6, 16, 151

Fahrenheit temperature, 120
Faraday, Michael, xv, 77, 79–80
feedbacks, 113
  climate, 113
  and global climate modeling, 140
  involving clouds, 10, 122, 141
  involving ice and snow, 115, 137
  involving sea level rise, 116
  involving water vapor, 114
  positive, 114–115
Foote, Eunice Newton, 47, 85–86
forecasting, weather, xi, 110, 123, 126, 131, 142
fossil fuels, xiii, 21, 44, 92
Fourier, Jean-Baptiste Joseph, 77, 80
Fourth Assessment Report of IPCC, 4, 5, 143
Framework Convention on Climate Change, 3, 5

Gandhi, Mahatma, 56, 154, 158
Gates, Bill, 50
Gautier, Catherine, 20–21
GCMs (general circulation models or global climate models), 111
gedanken experiments, 108, 136, 142

geoengineering, 37
global warming, x, 6, 61
Global Weather Experiment, 129
Goddard Institute for Space Studies, xi, 10
Gore, Al, 3–5, 33
GRACE satellite, 35, 117
greenhouse effect, 17
  described, 77
  as a global problem, 88
  human-caused increase in, 60, 86, 94, 142
  long-term results of changes in, 88
  research on, 78, 82
  role of clouds in, 83, 117
  role of water vapor in, 114
  theoretical research on, 88
  validity of, 72
greenhouse gases, 77, 83, 100
Greenland, 31, 36, 93, 116

Hamilton, Louisa, 80
Hassol, Susan Joy, 55
heat pump system, 43, 45
Herschel, William, 80, 85
historical record of atmospheric concentrations, 93
horizontal resolution, increasing, 110
humidity advection, 106
humidity gradient, 106
hurricanes, 61, 62, 112, 134

ice ages, 24, 36, 73, 81, 116
identical-twin experiment, 128
infrared radiation, 77–78, 80, 82, 85, 92, 118
initial conditions, 105, 126, 128
initial-value problem, 132–133
Institute for Advanced Study, 107
intelligence, artificial, 142
Intergovernmental Panel on Climate Change, 5, 11
IPCC assessment reports, 11

jargon, ix, 67

Kasich, John, 21
Keeling, Charles David, xii–xiv, 89
Keeling, Ralph, 90
Keeling curve, xiii, 65, 89–91
Kelvin, Lord (William Thomson), 119, 139
Kelvin temperature, 117–118
Kerry, John, 33
King, Jr., Martin Luther, 56, 154

Index

light, visible, 83, 117
longwave radiation, 85
Lorenz, Edward, 126–127

Mandela, Nelson, 56, 154, 158
Manhattan Project, 1
Mars, 84, 122
massively parallel computation, 107
mathematical models, 104, 108
Mauna Loa, xiii
McNutt, Marcia, 22
medical metaphors, 33, 38
meteorology, xi, 47–48, 104, 107
methane, 17, 77, 86, 93–94, 100
microinverters, 45
Montreal Protocol, 15

National Center for Atmospheric Research, xi, 10
negative emissions, 37
nitrogen, 83, 92
nitrous oxide, 17, 77, 94, 100
Nobel Prize, 4, 6, 81
nuclear power, 69–70, 150, 154
numerical analysis, 104–105, 107
numerical weather prediction, xiv, 107, 111

observational evidence for global climate change, 31, 34, 57–58, 72, 150
oceans
  becoming more acidic, 31
  and climate models, 140
  containing most of heat added to climate, 31, 34
  exchanging carbon with atmosphere, 94
  observed by Argo program, 34
  returning carbon to, xii
  rise in sea level, 116
  rise in temperature of, 62, 114
Oeschger, Hans, 15, 17
ozone
  and absorption of ultraviolet radiation, 77
  and destruction by chlorofluorocarbons (CFCs), 15
  hole, 6–7, 15, 63
  layer of in stratosphere, 5
  role in the greenhouse effect, 17, 77, 94

Paris Agreement, 20, 22, 27, 145, 154, 156
Pennsylvania State University (Penn State), 47–48

pH, 36
photosynthesis, 94
Piketty, Thomas, 48
precipitation, 44, 116, 138
predictability of climate, 131–132
predictability of weather, 123, 131–132
predictability theory, 130
prediction, weather. *See* weather forecasts

radiant heat, 80, 118
radiation, 92, 93, 117, 120
radiative signature, 93
rainforests, 91–92, 109
Ramanathan, Veerabhadran, 56, 120
relative humidity, 92
renewables, 26, 70, 157
Republican party, 40, 71, 153
Revelle, Roger, xii–xiii, 3
Richardson, Lewis Fry, 103, 106–107, 110
Rowland, F. Sherwood, 6
Royal Institution of Great Britain, 77, 79, 83
Royal Society, 79

sampling error, 96–97
Schneider, Stephen H., 9, 10
Scripps Institution of Oceanography, xii–xiv, 3, 89–90, 156, 159
sea ice, loss of, 16, 73, 115, 137, 151
sea level rise, ix, 22, 35, 116–117, 145, 151
sensitive dependence on initial conditions, 126
shortwave radiation, 85
Siegenthaler, Ulrich, 15, 17
Sixth Assessment Report of IPCC, ix, 143, 145
Smil, Vaclav, 44, 46, 48, 51, 157
solar energy, 43–45
solar modules, 43, 45, 50
solar photovoltaic system, 43, 45–46
solar radiation, 77, 83, 117–118
soot, 17, 94
species extinction, 57
steroids, 39
Suess, Hans, xii
Summary for Policymakers of IPCC, 7, 143–145
supercomputers, 107, 130

temperature
  and 2 degrees Celsius warming target, 14, 20, 36–37, 145
  atmospheric, 107
  and climate modeling, 130, 136

temperature (cont.)
   global average temperature, changes in, 14, 18, 21, 140, 144, 155
   global average temperature of oceans, 96
   historical record of, 34, 95
   as one of many climate aspects, 5, 61, 101
   reliability of data, 95–97
   variability of, 26, 48, 98, 102
temperature advection, 106
terrestrial radiation, 117–119
thermostat, 22, 27, 38, 65, 137
thought experiments, 108, 110
Truman, Harry S., 1
Tyndall, John, 47, 77, 80, 82, 85

ultraviolet radiation, 77
Uncle Pete, 23–24, 28, 60, 71

United Nations, 3, 15, 20, 144, 156
University of California, San Diego, x, xi, 4, 89
urban heat island effect, 97
urgency, 13, 19, 154, 156, 161

Venus, 84, 122
volcanoes, 24–25, 101, 159

warmest years on record, 31, 34, 57, 95
warming trend, 95, 150
water vapor, 17, 34, 77–78, 81, 92, 114
weather forecasts, 104, 107, 110, 123–124, 129, 130
wind power, 70
Working Group One of IPCC, ix, 30, 143, 145
World Meteorological Organization, 155

For EU product safety concerns, contact us at Calle de José Abascal, 56–1°, 28003 Madrid, Spain or eugpsr@cambridge.org.

www.ingramcontent.com/pod-product-compliance
Lightning Source LLC
LaVergne TN
LVHW021713060526
838200LV00050B/2634